Hydraulic performance of an impermeable submerged structure for tsunami damping

submitted to and approved by

the Department of Architecture, Civil Engineering and Environmental Science
of Technische Universität Carolo-Wilhelmina
zu Braunschweig

in candidacy for the degree of
Doktor - Ingenieur (Dr.-Ing.)

Dissertation

by

Agnieszka Strusińska
born 3rd Februar 1980
from Szczecin, Poland

Submitted: 14th Januar 2010
Oral examination: 23rd April 2010

Examiners: Prof. Hocine Oumeraci
Prof. Holger Schüttrumpf

2010

Dissertation

submitted to and approved by the Department of Architecture, Civil Engineering and Environmental Science of Technische Universität Carolo-Wilhelmina zu Braunschweig
in candidacy for the degree of
Doktor - Ingenieur (Dr.-Ing.)

Submitted: 14th Januar 2010
Oral examination: 23rd April 2010
Examiners: Prof. Hocine Oumeraci
Prof. Holger Schüttrumpf

Agnieszka Strusińska

HYDRAULIC PERFORMANCE OF AN IMPERMEABLE SUBMERGED STRUCTURE FOR TSUNAMI DAMPING

ibidem-Verlag
Stuttgart

Bibliografische Information der Deutschen Nationalbibliothek
Die Deutsche Nationalbibliothek verzeichnet diese Publikation in der Deutschen Nationalbibliografie; detaillierte bibliografische Daten sind im Internet über http://dnb.d-nb.de abrufbar.

Bibliographic information published by the Deutsche Nationalbibliothek
Die Deutsche Nationalbibliothek lists this publication in the Deutsche Nationalbibliografie; detailed bibliographic data are available in the Internet at http://dnb.d-nb.de.

∞

Gedruckt auf alterungsbeständigem, säurefreien Papier
Printed on acid-free paper

ISBN-13: 978-3-8382-0212-9

Printed in Germany

To my family

Acknowledgements

I would like to express my greatest gratitude to the people whose wisdom, knowledge and experience contributed to the accomplishment of this thesis:

to prof. Oumeraci for his constant support, professional advice and criticism of my study as well as encouragement during the last phase of the research.

to prof. H. Oumeraci, prof. D. Dinkler, Dr. A. Kortenhaus and Dr. U. Kopka for providing me the opportunity both to conduct this research, to collect the experience in the laboratory of LWI, to attend and present the results of my work at local and international conferences, and their scientific supervision.

to prof. H. Schüttrumpf, prof. H. Oumeraci, prof. A. Dittrich and prof. D. Dinkler for critical revision of my thesis and pleasant atmosphere during my examination.

to Dr. P. Lynett for his incredible interest in my research, scientific support, patience and understanding not only during my stay at the A&M University in College Station, but also during my whole research period in Braunschweig; for the opportunity of both using his numerical model COULWAVE and to improve my skills as a beginning numerical modeller.

to prof. P.L.-F. Liu, prof. S. Grilli, prof. P.A. Madsen and prof. V. Titov for their great help in solving the faced scientific problems.

to Mr. J. Grüne and Mr. R. Schmidt-Koppenhagen, Dr. M. Matsuyama and Dr. T. Yasuda for making the results of their research available for the purposes of the numerical model validation.

to Dr. S. Manam for his honesty, great friendship, sharing daily activities as well as the assistance in the first year of my research.

to Mr. B. Lehmann and Mr. B. Neumann for the perfect design and construction of the artificial reef used in the laboratory experiments as well as their help and assistance during the performance of the tests.

to all my colleagues from LWI and Graduate College 432 for a kind reception, friendship and sharing the knowledge and experience. Special gratitude for Mrs. G. Fournier and Mrs. R. Bähr for their assistance in fixing the daily matters, for Dipl. Ing. M. Kudella and Dipl. Ing. M. Brühl for the introduction to conduction of laboratory experiments and their scientific help and for Mr. R. Kvapil for his warmness and great help in the laboratory.

to students U. Berndt, S. Koss and K. Alem for their great help, assistance, engagement and patience when performing the laboratory experiments in the wave flume of LWI.

to my family for the constant support, strong belief that I will finish the thesis one day and providing me the opportunity to achieve a high educational level, sometimes even by sacrificing their needs and dreams.

to Kamila and Grzegorz for encouraging me to apply for the PhD position at the Braunschweig University of Technology, their loyalty, friendship, hospitality and the feeling of being at home.

to my dear friends met during the short term research at the A&M University in College Station for their friendship and help in solving scientific problems. Special gratitude for my Iranian friends for a warm welcome into their group and giving me the opportunity to experience their rich culture and hospitality.

to myself finally, for the strength and motivation to finish the research despite the faced obstacles.

Abstract

Besides early-warning systems and further non-structural tsunami mitigation measures, the development of effective coastal defence structures against tsunami has become a hot research issue since the 2004 Indian Ocean Tsunami. A multi-defence line strategy aiming at the gradual attenuation of tsunami by a combination of both natural barriers and man-made structures has been proposed as one of the most promising alternatives. The first defence line in this strategy, an artificial reef, should damp the wave before reaching the shore. It is obvious that the functioning of the entire defence system will significantly depend on the hydraulic performance of this first defence line.

The main objective of this study is therefore to investigate through laboratory experiments and numerical modelling the local processes (wave breaking, wave fission) and the global processes (wave transmission, wave reflection and wave energy damping) associated with the interaction of solitary wave-like tsunami with an impermeable artificial reef as a first step to determine the required dimensions of the artificial reef and thus for a more systematic feasibility study.

An extensive review and analysis of the available knowledge has shown that: (i) the experience in damping of storm waves by using submerged structures cannot be directly applied to tsunami due to the different characteristics of these waves, (ii) there is no systematic study on the nonlinear transformation of a solitary-wave tsunami wave over submerged structures of finite widths.

The laboratory experiments in the 2 m wide flume of Leichtweiß-Institute for Hydraulic Engineering and Water Resources (LWI) for different impermeable reef geometries and solitary wave conditions aim at the classification of breaker types, the derivation of breaking criterion and the determination of the number of the solitons due to the fission process. The experiments have shown that: (i) the breaking criterion depends on the relative structure width and the relative incident wave height, (ii) the structure shape does not influence breaking conditions, (iii) the soliton numbers increase for smaller submergence depths and wider structures and is independent of the incident wave height for a given water depth over the structure.

The numerical investigations using the Boussinesq-type model COULWAVE, which is modified in this study to introduce the computation of solitary wave energy, have indicated that noticeable wave energy reduction is achieved only for waves breaking over the reef and is dependant on structure width and submergence depth.

Overall, this study has contributed to an improved understanding of the hydraulic performance of artificial reefs under tsunami impact and to identify the crucial knowledge gaps related to the nonlinear interaction of solitary waves with submerged structures of finite widths.

Kurzfassung

Seit dem verheerenden Tsunami 2004 im Indischen Ozean ist die Entwicklung von effektiven Küstenschutzmaßnahmen gegen Tsunamiereignisse, neben Frühwarnsystemen und weiteren nicht-konstruktiven Maßnahmen, zunehmend zu einem bedeutenden wissenschaftlichen Forschungsthema geworden. Hierbei stellt eine Multi-Schutzlinien-Strategie, die auf eine graduelle Dämpfung des Tsunamis durch eine Kombination aus natürlichen Barrieren und künstlichen Bauwerken abzielt, eine der vielversprechendsten Alternativen dar. Die erste Schutzlinie in dieser Strategie ist ein künstliches Riff im Küstenvorfeld zur Dämpfung der Wellen, bevor diese die Küste erreichen. Es ist dabei offensichtlich, dass die Effektivität des gesamten Schutzsystems ganz wesentlich von der hydraulischen Wirksamkeit dieser ersten Schutzlinie abhängt.

Das wesentliche Ziel dieser Arbeit ist daher, die lokalen Prozesse (Wellenbrechen, Aufspaltung der solitäre Wellen in Solitone) und die globalen Prozesse (Wellentransmission, Wellenreflexion und Wellendämpfung) sowohl durch Versuche im Wellenkanal als auch durch numerische Modellierung zu untersuchen. In einem ersten Schritt wird die Interaktion zwischen einem Tsunami (dargestellt durch eine solitäre Welle) und einem undurchlässigen, künstlichen Riff untersucht. Hierdurch lassen sich anschließend die erforderlichen Abmessungen des Riffs für eine vorgegebene Tsunamidämpfung bestimmen.

Eine intensive Analyse des vorhandenen Wissensstandes hat gezeigt, dass (i) die vorhandene Erfahrung auf dem Gebiet der Wellendämpfung bei Sturmfluten durch getauchte Strukturen nicht direkt auf Tsunami übertragen werden kann, (ii) keine systematischen Untersuchungen zur nichtlinearen Transformation einer solitären Welle über getauchte Strukturen endlicher Breite vorliegen.

Die notwendigen Versuche wurden mit undurchlässige Riffgeometrien und solitären Wellen im 2 m breiten Wellenkanal des Leichtweiß-Instituts für Wasserbau (LWI) durchgeführt. Die Versuche zielten auf die Klassifizierung von Brechertypen sowie auf die Herleitung von Brecherkriterien und die Bestimmung der Anzahl der Solitone ab, die durch den Aufspaltungsprozess der Welle am

Bauwerk entstehen. Die Versuche haben dabei gezeigt, dass (i) das Brecherkriterium von der relativen Bauwerksbreite und der relativen Wellenhöhe der einlaufenden Welle abhängt, (ii) die Form des Bauwerks keinen Einfluss auf das Brechverhalten hat und (iii) die Anzahl der Solitone bei kleiner werdender Eintauchtiefe und breiteren Strukturen ansteigt und unabhängig von der ankommenden Wellenhöhe bei konstanter Wassertiefe über dem Bauwerk ist.

Die numerischen Untersuchungen basieren auf dem Boussinesq-Modell COULWAVE, das im Rahmen dieser Untersuchungen modifiziert wurde, um die Berechnung der Wellenenergie der solitären Welle zu implementieren. Dies hat gezeigt, dass eine deutliche Reduktion der Wellenenergie nur dann erreicht werden kann, wenn die Wellen über dem Riff brechen. Die Energiereduktion ist somit von der Riffbreite und der Eintauchtiefe abhängig.

Insgesamt haben diese Untersuchungen zu einem verbesserten Wissenstand im Bezug auf die hydraulische Wirksamkeit von künstlichen Riffen unter Tsunamieinwirkung beigetragen. Dadurch konnten darüber hinaus die wesentlichen Wissenslücken aufgezeigt werden, die im Zusammenhang mit der nichtlinearen Interaktion von solitären Wellen mit getauchten Strukturen endlicher Breite stehen.

Table of contents

List of figures

List of tables

List of symbols

Capital Latin Letters

Symbol	Definition	Unit
A_1	parameter in solitary wave solution by Wei and Kirby (1995)	[m/s]
A_2	parameter in solitary wave solution by Wei and Kirby (1995)	[m^{-1}]
A_3	parameter in solitary wave solution by Wei and Kirby (1995)	[m^{-1}]
A_4	parameter in solitary wave solution by Wei and Kirby (1995)	[m]
B	width of structure crown	[m]
B_r	parameter controlling wave breaking in COULWAVE model	[-]
$B_{1,2,3}$	coefficients	[-]
E	wave energy	[J/m^2]
E_d	dissipated wave energy	[J/m^2]
E_i	incident wave energy	[J/m^2]
E_k	kinematic wave energy	[J/m^2]
E_o	wave energy in farfield	[J/m^2]
E_p	potential wave energy	[J/m^2]
$E_{p,w}$	potential energy due to the wave only	[J/m^2]
E_r	reflected wave energy	[J/m^2]
E_t	transmitted wave energy	[J/m^2]
E_{tot}	total wave energy	[J/m^2]
F_{br}	component of vector of energy dissipation due to wave breaking (along "x" axis)	[m/s^2]
G_{br}	component of vector of energy dissipation due to wave breaking (along "y" axis)	[m/s^2]
H	wave height	[m]
H_b	wave height at breaking	[m]
H_{crit}	critical wave height	[m]
H_d	dissipated wave height	[m]
H_i	incident wave height	[m]
$H_{i,1}$	incident wave height of solitary wave in own experiments, measured from still water level to wave crest	[m]
$H_{i,2}$	incident wave height of solitary wave in own experiments, measured according to Eq. (3.3)	[m]
$H_{i,3}$	incident wave height of solitary wave in own experiments, measured from wave trough at the rear part of the wave to wave crest	[m]

$H_{i,nom}$	nominal incident wave height	[m]
H_{m0}	zero momentum wave height	[m]
H_o	deep-water (offshore) wave height	[m]
H_r	reflected wave height	[m]
H_s	wave height in shallow water	[m]
H_t	transmitted wave height	[m]
$H_{1/3}$	significant wave height	[m]
K_d	wave energy dissipation coefficient	[-]
K_r	wave reflection coefficient	[-]
K_t	wave transmission coefficient	[-]
L	wavelength	[m]
L_i	incident wavelength	[m]
$L_{i,1}$	incident wavelength of solitary wave corresponding to incident wave height $H_{i,1}$	[m]
$L_{i,2}$	incident wavelength of solitary wave corresponding to incident wave height $H_{i,2}$	[m]
$L_{i,3}$	incident wavelength of solitary wave corresponding to incident wave height $H_{i,3}$	[m]
$L_{i,nom}$	nominal incident wavelength	[m]
L_o	deep-water (offshore) wavelength	[m]
L_r	reflected wavelength	[m]
L_t	transmitted wavelength	[m]
M	parameter used for calculation of soliton number in approach by Germain (1984) and Kabbaj (1985)	[-]
N	solitons number	[-]
$\mathbf{R}_{br}$	vector of energy dissipation due to wave breaking	[m/s^2]
$\mathbf{R}_{bf}$	vector of energy dissipation due to bottom friction	[m/s^2]
$\mathbf{R}_{sev}$	vector of eddy viscosity dissipation in the Smagorinsky-type subgrid model	[m/s^2]
T	wave period	[s]
T_i	incident wave period	[s]
T^*	transition time of breaking event	[s]
U	Ursell number	[-]
$\mathbf{U}$	velocity vector	[m/s]

Small Latin Letters

Symbol	Definition	Unit
a	wave amplitude	[m]
a_i	incident wave amplitude	[m]
a_o	deep-water (offshore) wave amplitude	[m]
b_l	width of a canal in Lamb`s (1932) approach	[m]

b_2	width of a canal in Lamb`s (1932) approach	[m]
c	wave celerity	[m/s]
c'	dimensionless wave celerity	[-]
c_{g1}	group velocity in canal of width b_1 in Lamb`s (1932) approach	[m/s]
c_{g2}	group velocity in canal of width b_2 in Lamb`s (1932) approach	[m/s]
c_o	deep-water (offshore) wave celerity	[m/s]
d_r	submergence depth	[m]
d_{50}	median grain diameter	[m]
$d_{50,a}$	armour diameter size	[m]
f	wave frequency	[Hz]
f_b	bottom friction coefficient	[-]
f_p	wave peak frequency	[Hz]
g	gravitational acceleration	[m/s^2]
h	water depth	[m]
h_b	water depth at breaking	[m]
h_r	height of submerged structure	[m]
h_o	offshore water depth	[m]
k	wave number	[m^{-1}]
	turbulent kinetic energy	[m^2/s^2]
k_1	wave number of bound waves in second harmonics	[m^{-1}]
k_2	wave number of free waves in second harmonics	[m^{-1}]
m	gradient of bottom slope ($m=\tan\alpha$)	[-]
	mass of water column	[kg]
m_o	zero momentum of energy spectrum	[m^2]
p	structure porosity	[%]
t	time	[s]
t_o	time of breaking event initiation	[s]
u	horizontal velocity along "x" axis	[m/s]
$\mathbf{u}$	horizontal velocity vector along	[m/s]
u_b	horizontal velocity evaluated at sea bottom	[m/s]
$\mathbf{u}_b$	horizontal velocity vector evaluated at sea bottom	[m/s]
u_α	horizontal velocity along "x" axis evaluated at water level z_α in the Bousinesq-type equations by Nwogu (1993)	[m/s]
$\mathbf{u}_\alpha$	horizontal velocity vector evaluated at water level z_α in the Bousinesq-type equations by Nwogu (1993)	[m/s]
$\bar{u}$	depth-averaged horizontal velocity along "x" axis	[m/s]
$\bar{\mathbf{u}}$	depth-averaged horizontal velocity vector	[m/s]
u_1	horizontal velocity along "x" axis in canal of width b_1 in Lamb`s (1932) approach	[m/s]
u_2	horizontal velocity along "x" axis in canal of width b_2 in Lamb`s (1932) approach	[m/s]
v	horizontal velocity along "y" axis	[m/s]

v_α	horizontal velocity along "y" axis evaluated at water level z_α in the Bousinesq-type equations by Nwogu (1993)	[m/s]
$\bar{v}$	depth-averaged horizontal velocity along "y" axis	[m/s]
w	vertical velocity along "z" axis	[m/s]
x	wave propagation distance measured along "x" axis	[m]
x_s	distance from submerged structure to a shore	[m]
x_1	length of shelf slope	[m]
x_2	length of region in front of artificial reef	[m]
x_3	total length of artificial reef	[m]
x_4	length of region behind artificial reef	[m]
z	water level	[m]
z_α	water level at which horizontal velocity vector is evaluated in the Bousinesq-type equations by Nwogu (1993)	[m]
$\bar{z}$	elevation of gravity centre of water column	[m]

Capital Greek Letters

Symbol	Definition	Unit
Δ	parameter used for calculation solitary wave surface elevation	[-]
Δt	temporal grid resolution	[s]
Δx	spatial grid resolution along "x" axis	[m]
Δz	spatial grid resolution along "z" axis	[m]

Small Greek Letters

Symbol	Definition	Unit
α	sea bottom slope	[°]
	coefficient used for calculation of initial solitary wave profile according to Wei and Kirby (1995) solution	[-]
α_1	seaward structure slope	[°]
α_2	landward structure slope	[°]
β	coefficient used for calculation of initial solitary wave profile according to Wei and Kirby (1995) solution	[-]
	surf similarity parameter by Yoo (1986)	[-]
γ_d	wave steepness breaking criterion	[-]
γ_s	water depth breaking criterion	[-]
δ_b	mixing length coefficient	[-]
ε	nonlinearity parameter	[-]
	turbulent dissipation rate	$[m^2/s^3]$
η	free surface water elevation	[m]
η_i	free surface water elevation of incident wave	[m]

η_r	free surface water elevation of reflected wave	[m]
η_t	free surface water elevation of transmitted wave	[m]
$\eta_t^{(I)}$	parameter representing initiation of breaking event	[m/s]
$\eta_t^{(F)}$	parameter representing cessation of breaking event	[m/s]
η_t^{*}	parameter representing initiation/cessation of breaking	[m/s]
λ_2	beat length of second harmonics	[m]
μ	frequency dispersion parameter	[-]
	dynamic viscosity	[Ns/m^2]
ξ	surf similarity parameter	[-]
ξ_b	surf similarity parameter as function of local wave properties	[-]
ξ_o	surf similarity parameter by Irribarren and Nogales (1949)	[-]
ξ_r	surf similarity parameter by Smith and Kraus (1991)	[-]
ξ_s	surf similarity parameter by Schüttrumpf (2001)	[-]
$\xi`$	surf similarity parameter by Battjes (1974)	[-]
$\xi*$	surf similarity parameter by Hara et al. (1992) for waves breaking over sloping beach	[-]
ξ_r*	surf similarity parameter by Hara et al. (1992) for waves breaking over rectangular structure	[-]
ξ_s*	surf similarity parameter by Hara et al. (1992) for waves breaking over submerged step	[-]
ξ_t*	surf similarity parameter by Hara et al. (1992) for waves breaking over trapezoidal structure	[-]
ρ	fluid density	[kg/m^3]
υ	eddy viscosity	[m^2/s]
ϕ	velocity potential	[m^2/s]
ω	angular frequency	[s^{-1}]

List of abbreviations

ARC	Above Reef Crown
Br	Breaking Waves
BBOS	Behind Beginning of Horizontal Part of Shelf
BBM	Benjamin, Bona, Mahony Equation
BEM	Boundary Element Method
BIEM	Boundary Integral Equation Method
BR	Behind Reef
BRST	Behind Seaward Ree Toe
Bq	Boussinesq-Type Equations
COBRAS	Cornell Breaking Wave and Structures
COLL.	Collapsing Breaker
COMCOT	Cornell Multi-Grid Coupled Tsunami Model
D	Diagrams Elaborated
DNS	Direct Numerical Simulation
E	Laboratory Experiments
F	Formulcessas Derived
FDM	Finite Difference Method
FM	Field Measurements
FVM	Finite Volume Method
gener.	Generated
inc.	Incident
I	Irregular Waves
Im	Impermeable Structure
KdV	Korteveg de Vries Equation
LES	Large Eddies Simulation
LRC	Landward Reef Corner
LRS	Landward Reef Slope
LRT	Landward Reef Toe
LWI	Leichtweiss-Institute for Hydraulic Engineering and Water Resources
LWT	Linear Wave Theory
MOST	Method of Splitting Tsunami
N	Numerical Study
NASA	National Areonautics and Space Administration
NB	Nonbreaking Waves
NSWE	Nonlinear Shallow Water Equations
ORC	Over Reef Crest

OS	Over Shelf
P	Permeable Structure
PLUNG.	Plunging Breaker
R	Regular Waves
RANS	Reynolds-Averaged-Navier-Stokes Equations
Rc	Rectangular Structure
RcS	Rectangular Step
rel.	Relative
T	Theoretical Study
Tr	Trapezoidal Structure
Trg	Triangular Structure
Tv	Thin Vertical Plate
S	Solitary Wave
SPILL.	Spilling Breaker
SURG.	Surging Breaker
SWL	Still Water Level
VOF	Volume of Fluid Method
WB	Wave Breaking
WF	Wave Fission
WG	Wave Gauge
1HD	One Horizontal Dimensional
2HD	Two Horizontal Dimensional
2D	Two Dimensional
3D	Three Dimensional
$\_^{m}$	Defined from measurements
$\_^{v}$	Defined from video records

1. Introduction

1.1 Position of the problem and motivations

Enormous devastation and large-scale socio-economic impact on coastal communities induced by extreme events such as the 2004 Sumatra Tsunami has revealed human unpreparedness in the face of this low-probability and high-consequence event. Such extreme hazards are hardly predictable and difficult to control. Due to the lack of an appropriate coastal protection against tsunami or the ineffectiveness of the existing mitigation measures, an urgent demand has appeared to develop novel and reliable defence systems. The defence strategy should be capable of providing a maximum efficiency even in such extreme conditions like those of the December 2004 Tsunami, which affected inter alia shorelines of Indonesia, Thailand, Sri Lanka, India, and partly Africa, causing more than 200,000 fatalities.

While the establishment of tsunami early warning systems and other non-structural countermeasures such as educational programmes or (re)settlement policy have attracted most of the undertaken efforts after the 2004 Indian Ocean Tsunami event, very little consideration has been directed towards structural means for a coastal protection. The tsunami defence structures play the primary role in situations, when there is not enough time for evacuation of local residents after issued tsunami warning as well as in the protection of facilities of strategic functions (e.g. power plants, water intakes, harbours).

Since a complete stoppage of a tsunami is neither possible nor economically affordable due to the large-sized protective structures, their negative influence on the surrounding environment and unpredictable effects in case of their failure, solutions based on the concept of a gradual dissipation of incoming tsunami energy would be required. Such a multi-defence lines strategy extending from offshore to far onshore has been proposed by Oumeraci (2006). It incorporates a combination of man-made structures (e.g. artificial reefs, sea walls) and natural barriers (e.g. coral reefs, dunes, mangroves), illustrated by Figure 1.1. The type and the number of the defence lines and barriers to be adopted in a specific region would depend on the local conditions, including the degree of exposure to tsunami hazard, already existing structural and natural mitigation measures, available construction materials, urbanization level, etc.

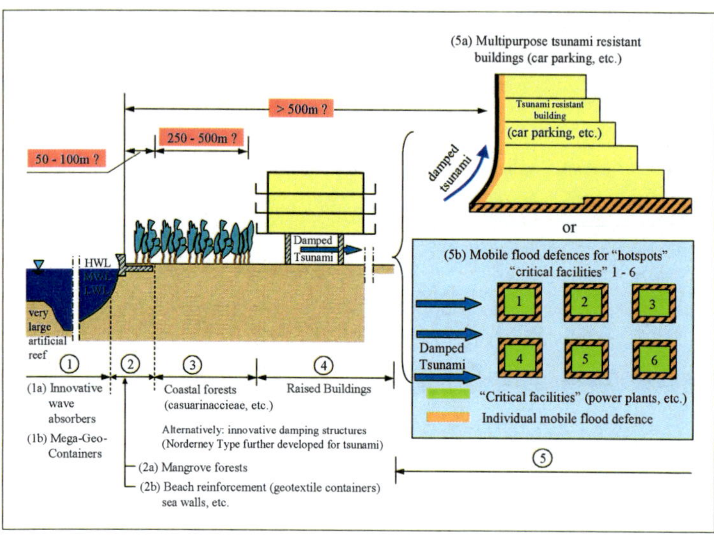

Figure 1.1: Multi-defence line strategy against tsunami (Oumeraci, 2006)

Moreover, the first defence line in the aforementioned strategy is believed to play the most decisive role for the overall performance of the entire protective system, since a significant rate of tsunami energy dissipation is expected to be induced by the defence structures located offshore. Such defence barriers are for instance low-cost submerged structures, called hereafter artificial reefs. The major advantage of the artificial reefs results from their broad and successful application to storm waves attenuation before the waves reach a coast. This effect is achieved by a reduction of wave height in the protected area essentially due to the forced, prematurely wave breaking over the structure crests. The same effect can be observed for their natural counterparts - coral reefs. In fact, the latter were reported to have significantly contributed to tsunami damping in the 2004 Sumatra event, although not yet confirmed by computational or experimental results.

However, large dimensions of an artificial reef will be expected in order to achieve a noticeable reduction of tsunami energy. Therefore, low-cost alternatives should be preferred, for example reefs made of geotextile mega-containers filled with dredged sand, as shown in Figure 1.2. A large experience is available to determine the dimensions of an artificial reef given a prescribed hydraulic performance in the case of the protection against storm waves. However, this experience cannot be readily extended for the defence against tsunami due to the significant differences between storm and tsunami waves (manifested mainly in wave period and carried energy).

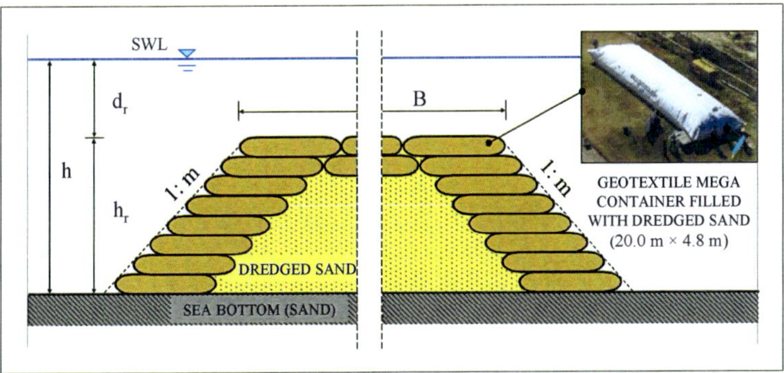

Figure 1.2: Proposed construction solution of artificial reef employed as protection against tsunami waves

1.2 Objectives

The primary objective of this thesis is therefore to examine the hydraulic performance of an artificial reef, leading to the specification of the structure dimensions required for effective tsunami energy attenuation. For this purpose, it would be necessary: (i) to identify the processes responsible for the hydraulic functioning of artificial reefs, (ii) to recognize the fundamental differences between tsunami and storm waves properties, and based on this knowledge, (iii) to use an improved available numerical model to predict the performance of the structure in terms of tsunami transmission, reflection, and energy dissipation.

1.3 Methodology

The practical limitations of tsunami modelling under laboratory conditions result from the vertical-horizontal aspect ratio of these long waves which is difficult to reproduce in a physical tank and cannot be expected to be fully overcome in the near future. The complexity of the mathematical models to describe long wave propagation from a deeper portion of water towards a coast where the nonlinear wave transformation takes place, excludes purely analytical solutions. Therefore, the analysis of the hydraulic functioning of impermeable artificial reef will be performed primarily by means of an available numerical model for water wave propagation.

The procedure adopted in this thesis is tentatively summarized in Figure 1.3. The first stage of this study (Chapter 2) is a review and analysis of the present state of knowledge including: (i) understanding of the tsunami phenomenon in comparison to the properties of wind-generated waves, (ii) selection of the appropriate wave theory governing tsunami propagation and its nearshore transformation, (iii) analysis of the existing theoretical, experimental and numerical models employed for the investigations on storm waves attenuation by means of low-crested structures, (iv) identification of the phenomena induced at submerged barriers with a particular consideration of the processes that are crucial for the tsunami attenuation. The objective of this stage of the study is to select an appropriate numerical model that could be further adapted to the formulated problem of tsunami damping by an artificial reef.

Chapter 3 provides description of own laboratory experiments which aimed at closing the existing gaps in the present knowledge on the nonlinear processes occurring at a submerged, impermeable structure of a finite width under tsunami-like solitary wave conditions.

In Chapter 4, the main features of the selected numerical source model for water wave propagation are presented. Validation of the model was performed on the basis of both available laboratory data and own experiments in order to ensure the correctness of the obtained numerical results.

In Chapter 5, numerical results of the extensive study on the hydraulic functioning of an artificial reef are discussed. The effectiveness of the structure to attenuate tsunami is defined for varying incident wave conditions and different barrier geometry by the rate of wave transmission, wave reflection and wave energy dissipation. As a final result, the structure dimensions corresponding to the most efficient tsunami damping are determined.

Chapter 6 contains the conclusions on the applicability and feasibility of artificial reefs acting as a coastal protection against tsunami waves. Also, the recommendations for future research are highlighted.

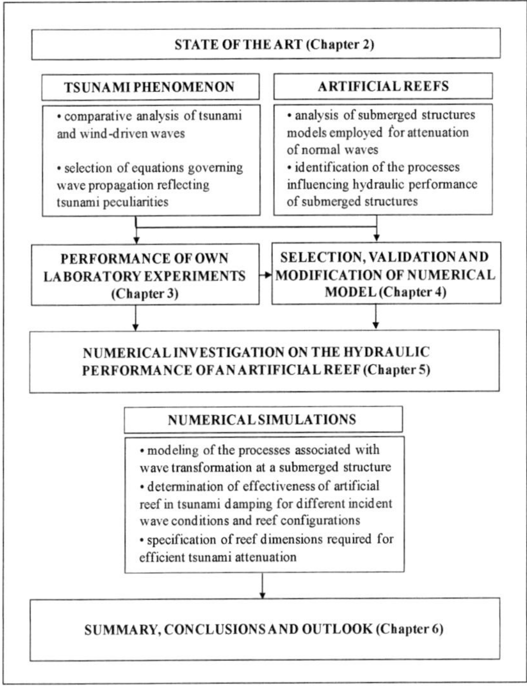

Figure 1.3: Methodology adopted in this study

2. Wave damping performance of submerged structures: state of the art review and applications for this study

The analysis of the issues covered by the literature study aims at examining the possibility of adapting the experiences available on the attenuation of storm waves by low-crested structures to tsunami damping and at the selection of an appropriate numerical model to perform the investigation. Therefore, a particular attention is directed towards the characteristics of a tsunami compared to storm waves as well as towards the processes associated with the hydraulic functioning of submerged structures and their numerical modelling:

- Description of the successive stages of tsunami evolution in comparison to the behaviour of wind-generated waves (Section 2.1).
- Discussion of the assumptions and concepts of water wave theories that are most appropriate to predict the long-wave tsunami phenomenon in coastal regions (Section 2.2).
- Analysis of hydraulic performance of low-crested barriers for different types of waves (Section 2.3).
- Summary of the main features of the available numerical models for water wave propagation and their application to the problem investigated in this study (Section 2.4).
- Specification of the objectives and methodology of the thesis that was tentatively formulated in the introductory chapter (Section 2.5).

More detailed information on each of the above listed problems can be found in Strusińska and Oumeraci (2009a, b).

2.1 Brief comparison of tsunami and storm waves

The identification of the most relevant differences/similarities between a tsunami and storm waves is required to explore the possibility to extend the available experience in the artificial reef design to the tsunami case, since the hydraulic performance of various types of submerged structures has only been investigated under storm wave conditions. Tsunami properties for each of the successive phases of its evolution from deep to shallow water will be emphasized in this section to asses the applicability of water wave theories to the considered stages of tsunami propagation.

2.1.1 Tsunami and storm wave characteristics

Tsunami is identified as a series of extremely long gravity waves travelling at very high speed in deep open ocean. In contrast to storm wave, which are induced by wind, a tsunami results from a sudden, large-scale, vertical disturbance of a water body, caused more often by seismic events (i.e. underwater shallow-focused earthquakes of minimum magnitude M_w 6.5 on the Richter scale), and less frequently by submarine landslides, volcanic activity or meteorite impacts (Camfield, 1980; Bryant, 1994; Ward, 2004).

Regardless the type of the disturbing forces, tsunami evolution consists of processes that are common for all types of water waves: generation, propagation from deep to shallow water, nearshore transformation (including wave refraction, reflection, diffraction, shoaling, breaking), and finally, wave run-up. However, the pattern, scale and intensity of these processes vary significantly due to the fundamental differences between tsunami and storm waves, manifested in the magnitude of wave period, phase speed, wave height and the amount of energy carried in the wave. The magnitude of these parameters is at first dependent on the characteristic of the source mechanism, and as waves approach a shore, it is additionally affected by local bathymetry, combined nearshore processes and coastline configuration.

A tsunami is generated as a series of waves spreading radially from the source once the tremendous amount of potential energy, released by the deformation of the seafloor during a submarine earthquake, is imparted into the overlying water column. The initial tsunami size and number of waves in the tsunami wave train are determined by the earthquake parameters (such as the magnitude and depth at which it occurs, fault geometry, etc.). The initial tsunami height can be approximated through the assumption that the initial surface profile mimics the sea bottom deformation. On the other hand, the storm waves observed normally on a beach, result from water surface disturbances caused by wind blowing over a sea, while a tsunami involves a motion of the entire water column stretched from the ocean bottom to the surface. The fact that the seismic energy transported with tsunami waves extends over the whole depth of the ocean basin and is not much dissipated during propagation, accounts for the devastating nature of a tsunami. Therefore, a tsunami can travel across an ocean, covering great distances from its origin, and being still hazardous when approaching a coast due to the insignificant energy losses.

In deeper ocean, tsunami height rarely exceeds 1.0 m. The typical tsunami period ranges between few minutes and one hour, what corresponds to the tsunami wavelength of the order of 100-500 km in deeper water (see Table 2.1). Since the offshore tsunami wavelength L_o is much greater compared to the local water depth h_o (i.e. $h_o/L_o<0.05$), the water wave celerity c_o near the source can be expressed according to the linear shallow-water wave theory as a function of the offshore water depth h_o only:

$$c_o = \sqrt{gh_o} \tag{2.1}$$

where g denotes the gravitational acceleration. The above equation clearly indicates that the deeper the water, the faster a tsunami propagates, approaching speed of 500-1000 km/h that is comparable to a speed of a commercial jet airplane.

As a tsunami enters coastal water, the wave speed slows down to 30-50 km/h because of decreasing water depth. Tsunami wavelength is also reduced, however wave period remains unchanged (see Table 2.1). The energy flux which is proportional to $a^2c = a^2\sqrt{gh}$ remains constant (energy loss are negligible small) and the decrease of wave speed in shallow water ($c = \sqrt{gh}$) will cause wave amplitude a to strongly increase by a factor of five and more. In addition to this shoaling effect, further effects due to refraction, diffraction and reflection may result in an even larger amplification of tsunami height onshore.

The storm waves are classified as short waves of the typical period of the order of 10 s and wavelengths reaching 300 m in deeper water (see Table 2.1). Their heights are usually less than 3 m and the waves are expected to shoal by a factor up to about two, depending on the steepness of the sea bed and the incident waves.

Table 2.1: Tsunami and storm wave properties in deep and shallow water, respectively

PARAMETERS		TSUNAMI WAVES		STORM WAVES	
DESCRIPTION	**UNIT**	**DEEPER WATER**	**COASTAL WATER**	**DEEPER WATER**	**COASTAL WATER**
Wave height	[m]	~ 1.0	10 – 15	5 – 15	1 – 5
Wavelength	[km]	100 – 500	10 – 20	0.16 – 0.62	0.07 – 0.19
Wave period	[s]	300 – 3600	300 – 3600	10 – 20	10 – 20
Phase speed	[km/h]	500 – 1000	30 – 50	56.2 – 112.5	25.2 – 35.7
Shoaling coefficient	[-]	Up to factor of 5-6		Up to factor of 2	
Frequency dispersion	[-]	Not dispersive		Very dispersive	

Since the velocity of short waves depends on wave frequency, the effect of frequency dispersion, which intensifies with the time/distance of wave propagation, should be taken into account for storm waves. As a result, waves start to disintegrate according to their frequency, with longer wave components travelling faster than the shorter ones.

Unlike storm waves, tsunami waves are generally not frequency dispersive and resemble solitary waves. In case of a local tsunami, a wave approaching a shore as a single input would be expected, whereas waves followed by numerous oscillations would be observed in case of a teletsunami reaching coasts located far from the generation source.

2.1.2 Implications for this study

The identified tsunami peculiarities can be briefly summarized as follows:

- Hardly predictable probability of tsunami occurrence.
- Great wavelengths corresponding to great tsunami periods.
- Small wave heights in deeper open ocean, yet strong shoaling nearshore.
- Extreme amount of energy contained in a wave, stretched over the whole water column.
- Small energy losses and almost no frequency dispersion during propagation.

2.2 Water wave theories

Accurate numerical modelling of the complex nearshore wave transformation would require a water wave model that accounts for both amplitude and frequency dispersion effects, predicting correctly all the processes for short, long, and intermediate waves. The accuracy and application of a numerical code is predominantly dependent on the limitations of the mathematical model selected to describe the wave motion. The assumptions and applications of the available water wave theories will be discussed in this section, with a particular consideration of those suitable for the prediction of tsunami propagation nearshore and interaction with submerged obstacles.

2.2.1 Available wave theories and assumptions

The most accurate mathematical model is ideally provided by a complete system of equations governing fluid motion, i.e. the *Navier-Stokes equations*, which can predict the nearshore processes for waves of any lengths, since the amplitude and frequency dispersion as well as energy dissipation are naturally incorporated. However, great costs are required for the computations. This is particularly due to the very fine spatial and time grids required to resolve energy dissipation (e.g. wave breaking, structure-induced turbulence, etc.), which is modelled at very high Reynolds number. Therefore, application of these equations to large-scale simulations of coastal hydrodynamics, such as tsunami propagation from deep to shallow water, is currently not feasible (Liu and Losada, 2002).

The introduced simplifications of the Navier-Stokes equations are based on the assumption, that the short-period and small-scale components of fluid motion can be neglected. This is achieved by averaging the flow properties over time or over space (Svendsen, 2006). As a result of the averaging of the Navier-Stokes equations over a typical time-scale of turbulence, the classical *Reynolds equations* are obtained. The tool employed to derive the Reynolds model from the Navier-Stokes equations is the Reynolds decomposition, which separates the flow variables into the mean (time-averaged) and fluctuating (turbulent) components. The additional term accounting for Reynolds stresses appears as a consequence of the unresolved turbulent fluid motion at the small time scale. Averaging of the Navier-Stokes equations over space leads to the numerical method termed *Large*

Eddies Simulation (LES), in which the unresolved fluid motion at small scales is represented by small eddies. However, both models still require much computational effort and thus cannot be applied to large-scale coastal problems.

Further assumptions on inviscid and irrotational flow provide a basis for the derivation of various water wave theories, which originate from the same fundamental hydrodynamic equations and the associated boundary conditions. The limitations of these mathematical models are essentially determined through the assumptions on the magnitude of the nonlinear and frequency dispersion effects. The resultant wave theories can be classified as follows:

- linear (Airy) wave theories:
  - deep-water wave theory or linear dispersive wave theory (small-amplitude, short waves),
  - intermediate-water wave theory (small-amplitude, moderate waves),
  - shallow-water wave theory or linear nondispersive wave theory (small-amplitude, long waves),
- nonlinear wave theories:
  - Stoke's wave theory (small-amplitude, short and moderate waves),
  - Boussinesq (1872), Boussinesq-types and Korteveg-de Vries (1895) (KdV) equations (moderate-amplitude, long waves),
  - cnoidal and solitary wave theories as the solution of the Boussinesq equation valid for one direction of wave propagation,
  - nonlinear shallow water wave equations (NSWE) (large-amplitude, long waves).

2.2.2 Nonlinear and dispersion effects

The relative importance of the effects of the amplitude and frequency dispersion for the distinction between the available wave theories was postulated by Ursell (1953). The *nonlinear effects* are defined by nonlinearity parameter ε, expressed as a ratio of wave amplitude a to water depth h (Svendsen, 2006):

$$\varepsilon = a / h \tag{2.2}$$

The profile of the nonlinear waves becomes asymmetric due to the fact that wave celerity depends on the wave amplitude. Therefore, steep wave crest propagates faster than wave troughs.

The measure of the *frequency dispersion* effects is dispersion parameter μ, formulated as a water depth-to-wavelength ratio:

$$\mu = h/L \tag{2.3}$$

As a result of the frequency dispersion phenomenon, wave separates into numerous linear components travelling at different speeds. Since the wave celerity is determined by frequency, not by amplitude, longer wave components propagate faster than their shorter counterparts.

By combining the nonlinear and the frequency dispersion effects, the Ursell number is obtained:

$$U = \frac{\varepsilon}{\mu^2} = \frac{a \cdot L^2}{h^3} \tag{2.4}$$

The value of the Ursell number reflects the application domain of the above mentioned mathematical wave models:

- $U << 0(1)$ linear (Airy) wave theory,
- $U << 10$ Stoke's wave theory,
- $U = 0(1)$ Boussinesq wave theory and KdV equations,
- $U >> 0(1)$ nonlinear shallow water (NSW) wave theory.

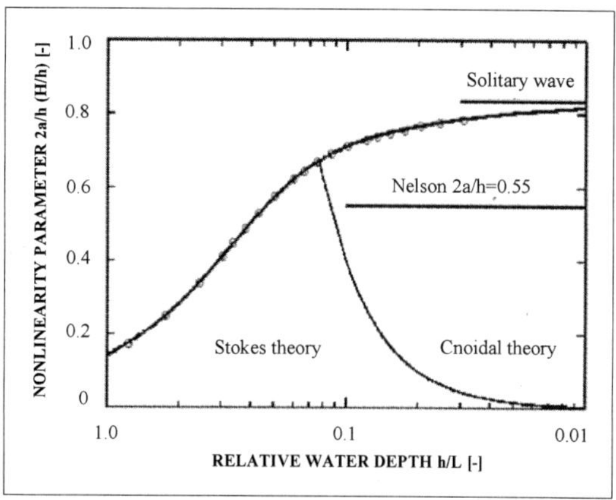

Figure 2.1: Application range of water wave theories (after Fenton, 2007)

Application of the aforementioned wave theories based on the nonlinear and frequency parameters is illustrated by Figure 2.1.

While the intensification of nonlinear effects in the coastal zone results directly from the nonlinear character of the nearshore processes, the simultaneous incorporation of the frequency dispersion effects (normally important for the deep water wave characteristics) is conditioned by the generation of shorter wave components through nonlinear interactions.

2.2.3 Applicability of wave theories for tsunami

a) Linear wave theory

The simplest water wave model corresponds to the *linear (Airy) wave theory*, which assumes an infinitesimal wave height compared to both water depth and wavelength (Svendsen, 2006). This clearly indicates that the nonlinear effects can be neglected, since $H/h \ll 1$ and that the wave steepness is also considered to be small ($H/L \ll 1$). With a very small Ursell number $U \ll 0(1)$, the linear wave theory is restricted to waves of small amplitudes and periodic motion, with their shape approximated by a sine/cosine curve. Primarily due to the closed orbit trajectory of water particle motion, resulting in no mass transport, the theory is poorly suitable to model progressive long waves. Within the framework of the Airy theory, there are three cases to be distinguished, depending on the ratio h/L:

- $h/L \leq 0.05$ linear shallow-water wave theory (nondispersive waves),
- $0.05 < h/L < 0.5$ linear intermediate-water wave theory (weakly dispersive waves),
- $h/L \geq 0.5$ linear deep-water theory (dispersive waves).

Considering the fact that a tsunami is a long wave, the linear shallow-water wave theory is generally accurate enough to describe tsunami propagation in deep ocean near its generation source, where both requirements for $H/h \ll 0(1)$ and $H/L \ll 0(1)$ are widely fulfilled.

b) Stokes wave theory

The requirement of a small wave height in comparison to water depth $H \ll h$ corresponds to the Stoke's wave theory, in which wave steepness H/L is additionally considered to be still relatively small, yet not infinitesimal small like in the case of linear waves (Svendsen, 2006). However, domination of the frequency dispersion effects h/L=0(1) indicates the limitation of this theory to short and intermediate waves, what immediately excludes Stoke's wave theory from a further consideration as a candidate model for describing a tsunami.

c) Depth-averaged equations

The two other categories of the nonlinear wave theories represent a system of depth-integrated equations, in which certain assumptions on the vertical profile of the flow field (horizontal and vertical velocity components, pressure) are made. As a consequence of the simplification of a three dimensional problem to a 2HD case by the elimination of the vertical dimension, a significant reduction of computational costs can be achieved (Svendsen, 2006). While the accuracy of these models is conditioned by the specified vertical profile of the aforementioned properties, their application to the long wave environment only is dictated by the solution approximation technique, in this case performed by means of long wave expansion or perturbation methods. The two categories are the nonlinear shallow water wave equations (NSWE) and the Boussinesq-type models.

Nonlinear shallow water equations (NSWE)

The simplest depth-integrated model, i.e. the nonlinear shallow water equations (NSWE), can be derived from the incompressible, irrotational equations for fluid motion, by integrating them over water depth and applying expansion technique in terms of small parameter μ (Svendsen, 2006). Since no restrictions on the wave height are introduced, the model is appropriate to simulate nondispersive (i.e. very long) waves of a finite amplitude $\varepsilon=0(1)>>0(\mu^2)$, such as a tsunami or tides. The horizontal velocity component is assumed to be constant over the entire water depth, while the vertical velocity component is equal zero, what corresponds to hydrostatic pressure distribution. Motion of these waves is translatory, i.e. there is water mass transport under a progressive wave. Due to the purely amplitude-dispersive character of the equations, the wave profile is not constant, yet it evolves in time, becoming steeper, since wave crests travel faster than wave troughs.

Wave breaking in the nonlinear shallow water equations is introduced through the dissipative Lax-Wendroff numerical scheme, in which the steep wave front is frozen before it becomes vertical (what corresponds to the incipient wave breaking), so that a bore-like energy dissipation of a wave of a permanent form is generated (Brocchini and Dodd, 2008). Unlike the real process of breaking, the length of the frozen wave front is limited to several grid points only. Another weakness of this scheme is the poor prediction of the location of incipient wave breaking, since the position of the frozen front is expressed as a certain distance from the numerical offshore boundary.

The nonlinear shallow water equations can be applied to simulate propagation of a tsunami wave in very shallow water, where the nonlinear effects become important through wave shoaling and breaking. Wave breaking event is however predicted purely numerically. Due to their nondispersive character of the equations, the phenomenon of tsunami fission over a continental shelf (reported e.g. by Matsuyama et al., 2007) cannot be reproduced correctly.

Boussinesq (1972) and Boussinesq-types equations

At present, the most advanced shallow water wave propagation models are the depth-averaged Boussinesq-type equations, accounting for both nonlinear and frequency dispersion effects. In the classical form, however, these effects are considered to be weak and tend to balance each other, since their leading orders are of the same magnitude $0(\varepsilon)=0(\mu^2) <<1$, $U_r=0(1)$. These models are obtained by the same depth-averaging technique as in the case of nonlinear shallow water equations. However, the expansion is performed in terms of small parameters ε and/or μ.Unlike the nonlinear shallow water equations, the horizontal and vertical velocity profiles vary over water depth, and consequently, the hydrostatic pressure distribution becomes invalid in this case (Kirby, 2003; Svendsen, 2006).

The *classical Boussinesq* (1872) *equations* were derived under assumptions of 1HD wave motion in constant water depth, with horizontal velocity evaluated at the bottom. Various alternatives to these equations have been developed, depending on the choice of the horizontal velocity components, i.e. equations expressed in terms of depth-averaged velocity, velocity at the sea bottom or at the free surface (Dingemans, 1997; Svendsen, 2006). This choice has been found to have a significant effect on the wave behaviour, particularly in deep water, where the dispersion relation and the wave celerity of the models diverge significantly from the actual wave behaviour. The accuracy of these models, however, remains the same. The set of two dimensional equations introduced by Peregrine (1967), representing the original Boussinesq equations (1872) expressed in terms of the depth-averaged velocity and extended to varying water depth, posses the best linear dispersion characteristic among the above mentioned alternatives (kh~0.75, with wave number $k=2\pi/L$).

The classical Boussinesq equations have recently attracted great attention directed towards removal of the constraints on both weak frequency dispersion and weak nonlinearity, i.e. to extend their application from shallow water to intermediate and deep water at which the dispersion parameter is of the order of unity, as well as to simulate larger wave heights in very shallow water where the nonlinear effects dominate. These approaches led to the development of new systems of equations termed generally the *Boussinesq-type models*.

The techniques employed to optimize the dispersion properties of the basic model allowed to obtain the so-called *improved (extended) Boussinesq-type equations* of satisfactory intermediate-depth dispersion characteristics compared to those predicted by the linear wave theory. Exemplarily, Witting (1984) successfully employed the Padé approximants to improve the dispersion characteristic up to kh~3-6, while Madsen et al. (1991) extended the limitation to kh~3 by modifying the dispersion terms. Nwogu (1993) on the other hand introduced velocity at an arbitrary water depth as one of the dependent variables. The suggested optimal level $z_\alpha=-0.531h$ (i.e. the water level for which the best agreement between the inherent and linear dispersion characteristics is obtained) corresponds to the intermediate water depth conditions with kh~3.

Further, in contrast to the classical Boussinesq theory with $0(\mu^2)$ accuracy, in which the quadratic polynomial approximation of the vertical flow field was applied, fourth and even higher order polynomials were used in the so-called *high-order Boussinesq equations* of accuracy $0(\mu^4)$. Together with an excellent enhancement of the linear dispersion relation, the costs of computational solution of these models also increased due to the resultant higher order spatial differentiation. Such models were for instance developed by Madsen and Schäffer

(1998) or by Gobbi et al. (2000), who achieved a range of the validity of the equations up to kh~6 by using a fourth-order polynomial approximation.

Simultaneously, a new type of models, the so-called *highly-nonlinear Boussinesq equations*, has been developed. The removed restriction of weak nonlinearity allowed to adjust the models to very shallow water environment, in which the shoaling waves are typically characterized by a great wave height-to-depth ratio. Liu (1994) and Wei et al. (1995) combined the strong nonlinear wave characteristics with the Nwogu′s (1993) approach, obtaining independently a fully nonlinear and weakly dispersive system of equations.

The newest approaches however allow to obtain the same accuracy in both nonlinear and linear properties. The procedure presented by Agnon et al. (1999) results in the accurate description of dispersive nonlinear waves up to kh~6 by expressing the free surface and bottom boundary conditions in terms of horizontal and vertical velocities evaluated at still water level, while applying truncated series expansions to the Laplace equation. This method was further improved by Madsen et al. (2002, 2003), who expanded the Laplace solution around an arbitrary water level, achieving so far the best linear and nonlinear characteristics accurate up to kh~40, with corresponding velocity profiles up to kh~12.

Apart from Nwogu (1993) and Madsen et al. (2002, 2003), the vertical distribution of velocity field was further enhanced by Lynett (2002) by introducing a two-layer model capable of predicting the velocity profiles up to kh~6.

The successful application of these models to intermediate-shallow water zones contributed to their further enhancement making them appropriate for the wide range of applications to such coastal processes as wave breaking, refraction, diffraction, wave-current interaction or effect of porous bottom. The energy dissipation associated with wave breaking and bottom friction is introduced artificially by adding dissipative terms into the momentum equation.

The mechanisms triggering wave breaking (usually based on empirical observations such as wave steepness limit) and the energy dissipation mode must both be specified. In the eddy viscosity models by Zelt (1991), Wei et al. (1995), Kennedy et al. (2000) and Lynett (2002), the magnitude and variation in a phase of the viscosity term was calibrated with laboratory experiments. Brocchini et al. (1992) and Schäffer et al. (1993), however, used a concept of a surface roller for spilling breakers, which requires determination of the roller thickness at each point and the roller orientation, based on experimental works. Finally, more sophisticated vorticity-based models are available (Yoon and Liu, 1989), in which the fluid motion is decomposed into a potential (non-breaking) and rotational part (standing for the vorticity generated by wave breaking). The experimental data to be provided are the boundary conditions for vorticity along the roller surface, determined from the observations of a hydraulic jump.

The rapid improvement of both nonlinear and linear dispersion properties of the Boussinesq-type equations over the recent years allow to model evolution of both nondispersive and dispersive tsunami waves of weak to high nonlinearity. In contrast to the nonlinear shallow water equations, prediction of wave breaking and associated energy dissipation is based on real representation of these processes.

Solitary wave theory

The analytical solution of the Boussinesq (1872) equations, under the assumptions of U=0(1), horizontal sea bottom and 1HD propagation in one direction only, was obtained by Korteveg-de Vries (1895). The solution describes periodic waves of a constant form – termed cnoidal waves, whose profile is defined by an elliptic cn-function. If the length of these waves approaches infinite ($L \rightarrow \infty$), the resultant wave form is a solitary wave, which consists of a wave crest only, located above still water level (Svendsen, 2006).

The available solitary wave solutions describing surface elevation η and wave celerity c are summarized below (Dingemans, 1997):

$$\eta(x,t) = H \operatorname{sech}^2\left(\frac{x-ct}{\Delta}\right) \tag{2.5}$$

with c and Δ determined as follows:

- solitary wave solution of Boussinesq (1872) equation:

$$c = \sqrt{g(h+H)} \text{ and } \Delta = \sqrt{4h^3/3H} \tag{2.6}$$

- solitary wave solution of Korteveg-de Vries (1895) equation (KdV):

$$c = \left(1+\frac{H}{2h}\right)\sqrt{gh} \text{ and } \Delta = \sqrt{4h^3/3H} \tag{2.7}$$

- solitary wave solution of Benjamin, Bona and Mahony (1972) equation

(BBM):

$$c=\left(1+\frac{H}{2h}\right)\sqrt{gh}\ \text{and}\ \Delta=\sqrt{\frac{4h^3}{3H}\frac{c}{\sqrt{gh}}} \tag{2.8}$$

Wei and Kirby (1995) provided analytical solitary wave solution to the extended Boussinesq-type equations by Nwogu (1993), in which surface elevation and horizontal velocity are dependent on the water level z_α=-0.531h (selected for the evaluation of horizontal velocity in order to obtain the best agreement with linear dispersion characteristic).

The length of the solitary wave, although infinite by definition, can be approximated by the assumption that 95% of the water volume is contained within the following distance (Dean and Dalrymple, 1991):

$$L=2\frac{2.12h}{\sqrt{H/h}} \tag{2.9}$$

A solitary wave represents a good model of long waves such as a tsunami, not only because of the translatory motion under which mass transport occurs, but also due to their great impact force exerted on structures and large run-up (e.g. Goring, 1979; Losada et al., 1989; Grilli et al., 1994; Liu and Cheng, 2001; Grüne et al., 2006; Lynett, 2007).

2.2.4 Implications for this study

Assuming the typical ocean depth to be in the range of h=1-4 km, a tsunami in deep open ocean of amplitude of approximately a=1.0 m can be characterized by the following magnitudes of the nonlinearity, frequency dispersion and Ursell parameters (Liu, 2007):

- nonlinearity parameter: $\varepsilon=a/h\approx O\left(2.5\times10^{-4}-10^{-2}\right)$,
- dispersion parameter: $\mu=h/L\approx O\left(2.5\times10^{-4}-10^{-1}\right)$,
- Ursell parameter: $U=\varepsilon/\mu^2\approx O\left(2.5\times10^{-3}-10^{2}\right)$.

Therefore, linear shallow water (nondispersive) wave theory can be applied to predict tsunami propagation in deep ocean near the generation source, where the nonlinear effects can be neglected (see Figure 2.2). However, in coastal region the nonlinear processes become dominant due to a strong tsunami shoaling, suggesting the use of nonlinear wave theories. The nonlinear shallow water equations would be preferable in case of a nondispersive tsunami propagating in very shallow water. However, laboratory tests of Matsuyama et al. (2007) indicated that the effect of wave dispersion due to tsunami fission over a continental shelf should be additionally considered – in this case a fully nonlinear, dispersive Boussinesq model would be preferable. The aspect of application of wave theories to the wave-structure interaction will be discussed in Section 2.3.

Solitary wave theory can be used to represent a progressive tsunami both under laboratory conditions and in numerical simulations, as indicated by the available studies. In the future research however, tsunami modelling by means of undular bore should attract more attention, since it is believed to reflect the tsunami peculiarities at a shore more accurately than the solitary wave theory, as discussed by Madsen et al. (2008).

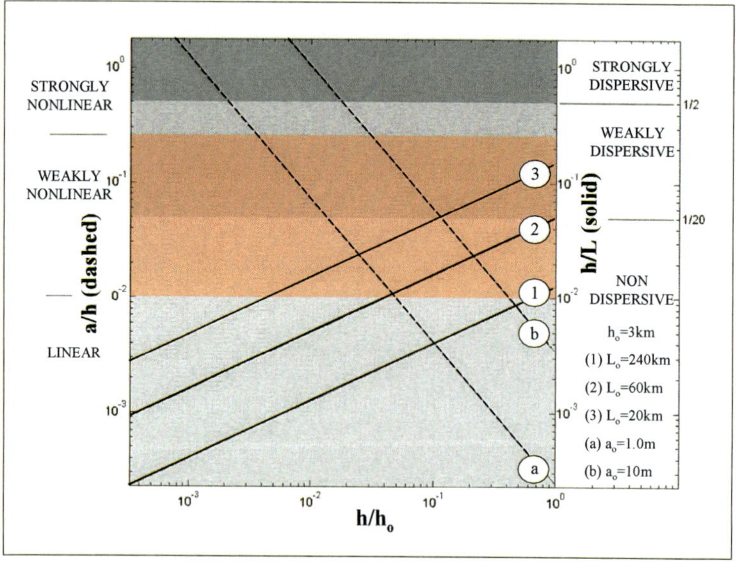

Figure 2.2: Nonlinear and frequency dispersion effects on tsunami behaviour at different water depths (Liu, 2007)

2.3 Hydraulic performance of low-crested structures

Due to the lack of plausible design guidelines for the application of an artificial reef as a coastal protection against a tsunami, understanding of the processes associated with the hydraulic performance of low-crested barriers, investigated primarily under storm wave conditions, is of utmost importance for this study. Particular attention will be put on: (i) the identification of the phenomena playing the crucial role for the damping performance of submerged structures, (ii) examination of the possibility to extend the already existing experience for normal waves to the case of a tsunami, (iii) recognition of the scientific gaps in the considered field that can be supplemented by performing of own laboratory experiments.

2.3.1 Classification of processes induced at submerged structures

According to the classification introduced by Bleck (2003), the phenomena associated with a wave passage over a low-crested structure are defined as *global* and *local processes*. The global processes lead to the changes of wave field in front of and behind the structure, which manifest themselves in the modification of the wave energy spectrum through: (i) wave energy reduction on the structure landside caused by different sources of energy dissipation, (ii) wave energy distribution, i.e. transfer into higher frequencies associated with the generation of nonlinear effects (higher harmonics in case of regular/irregular waves and solitons in case of a solitary wave). The processes belonging to this group are wave transmission, wave reflection and wave energy dissipation. The local processes should be considered as those induced locally, contributing to the generation of the global processes, namely wave breaking, nonlinear effects and vortex shedding (see Figure 2.3).

An incident wave approaching a low-crested structure shoals over the seaward slope, and at the same time being partially reflected back to the sea. Vortices at the upper and lower seaward corners of the obstacle are generated due to flow separation. The remaining part of the incident wave is transmitted behind the structure, possibly breaking over the structure crest, and due to the associated turbulent processes, significantly reducing wave energy. Some additional portion of the energy is dissipated due to friction at the reef crown and within the structure for permeable barriers. The shallower water over the barrier crest causes energy transfer into higher frequencies, what leads to the broadening of the transmitted energy spectrum. As a result, higher harmonics are generated in case of regular/irregular waves and solitons in case of a solitary wave. These short-period waves are then transmitted as free waves into the deeper part of the water behind the obstacle. Additionally, vortices at the landside of the structure are generated.

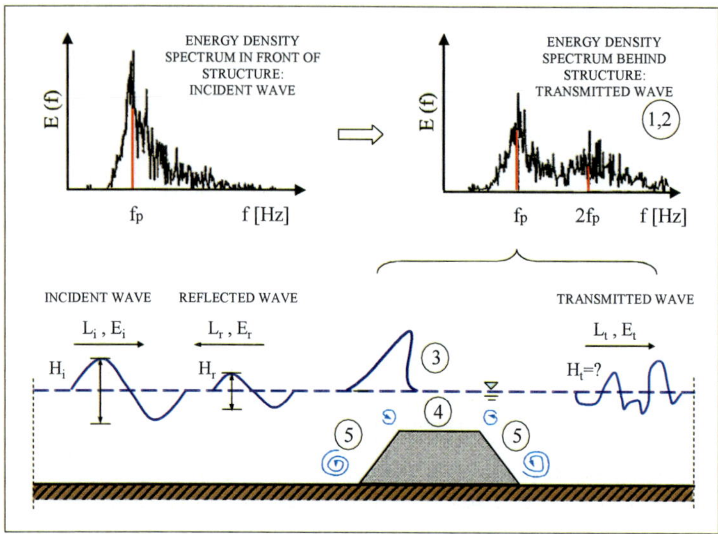

Figure 2.3: Definition of global and local processes: 1-wave energy dissipation, 2- energy transfer in wave spectrum, 3- wave breaking, 4- nonlinear effects, 5- vortex shedding (after Bleck, 2003)

2.3.2 Characteristics of global processes

a) Determination of energy coefficients

Global processes are governed by the *energy conservation law*, according to which the incident wave energy E_i is balanced by the sum of the transmitted E_t, reflected E_r and dissipated energy E_d:

$$E_t + E_r + E_d = E_i \tag{2.10}$$

The relationship expressed in terms of *energy coefficients* (transmission K_t, reflection K_r and dissipation K_d coefficients), can be employed to indicate the efficiency of a submerged structure to attenuate wave energy:

$$K_t^2 + K_r^2 + K_d^2 = 1 \tag{2.11}$$

The energy coefficients can be determined twofold:

- As a ratio of the transmitted H_t, reflected H_r, dissipated H_d wave height to the incident height H_i, using the proportionality between wave energy and wave height ($E \propto H^2$).

$$K_t = \frac{H_t}{H_i}, \quad K_r = \frac{H_r}{H_i}, \quad K_d = \frac{H_d}{H_i} \tag{2.12}$$

  Depending on the wave type and occurrence of nonlinear effects, the transmitted wave heights are determined as:

  - *Wave height measured from the wave trough to the wave crest* in case of transmission of regular and solitary waves, not accompanied by the nonlinear effects (e.g. Dattatri et al., 1978; Grilli et al., 1994).
  - *Significant wave height* $H_{1/3}$ in time domain analysis in case of transmission of irregular waves or transmission of regular waves accompanied by nonlinear effects. The significant height is determined as the mean of the largest 33% peak to trough heights of waves in the given time series (e.g. Powell and Allsop, 1985).
  - *Zero momentum wave height* H_{m0} in frequency domain analysis in case of transmission of irregular waves or transmission of regular waves accompanied by nonlinear effects (e.g. Ahrens, 1987; Bleck, 2003):

$$H_{m0} = 4\sqrt{m_0} \text{ where } m_0 = \int S(f)\ df \tag{2.13}$$

  with m_0 representing the zero momentum of the spectrum and $S(f)$ the energy density.

- Chang et al., 2001; Lin, 2004). The wave energy is computed as energy flux at a vertical cross sections located behind and in front of a submerged structure for the estimation of wave transmission and reflection, respectively.

$$K_t = \sqrt{\frac{E_t}{E_i}}, \quad K_r = \sqrt{\frac{E_r}{E_i}}, \quad K_d = \sqrt{\frac{E_d}{E_i}} \tag{2.14}$$

Unlike wave transmission K_t and reflection K_r coefficients, the rate of wave energy dissipation K_d cannot be measured directly and must therefore be calculated on the basis of the energy conservation relationship:

$$K_d = \sqrt{1 - K_t^2 + K_r^2} \tag{2.15}$$

The wave energy losses are essentially caused by turbulent effects induced by wave breaking, vortex shedding and bottom friction. Bleck (2003) reported that the energy losses due to generated vortices can reach approximately 18% of the overall rate of dissipated incident wave energy. Together with wave breaking, vortex generation are considered to be the most relevant sources of wave energy dissipation. The frictional losses over the artificial reef in comparison to energy reduction due to breaking are not significant and and can be generally neglected in case of smooth structure surfaces.

b) Parameters influencing global processes

The factors affecting the global processes can be generally classified into four main groups (see Figure 2.4): (i) incident wave conditions such as wave height H_i, wave period T_i or wavelength L_i, total water depth h, (ii) geometry of low-crested structure such as reef height h_r (or submergence depth d_r), crest width B, sea- and landside slopes $\tan\alpha_1$ and $\tan\alpha_2$, structure permeability (porosity p or armour diameter size $d_{50,a}$), distance from the shoreline x_s, (iii) underwater bathymetry such as sea bottom slope 1:m (where m=tanα), (iv) fluid properties such as fluid density ρ, dynamic viscosity μ, gravitational acceleration g.

However, these parameters expressed as dimensionless ratios are usually used in practice to determine the damping performance of a low-crested structure. The most common combinations of these parameters are: (i) relative incident wave height (H_i/h, H_i/d_r) indicating the magnitude of wave nonlinearity ε and representing wave breaking conditions over a submerged structure, (ii) incident wave steepness (H_i/L_i) which is important for wave breaking criteria in deep water, (iii) relative structure width (B/L_i, B/h) controlling wave breaking conditions over a submerged structure, (iv) relative submergence depth (d_r/h, d_r/H_i, d_r/L_i) which can be replaced by the relative structure height (h_r/h).

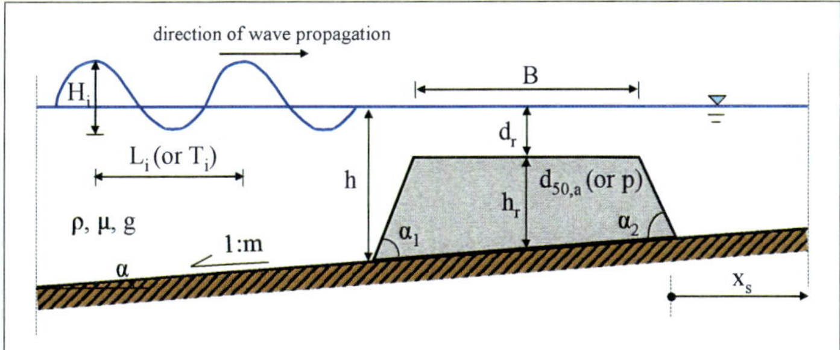

Figure 2.4: Definition of parameters affecting global processes

The effects of the most important parameters on the hydraulic performance of a submerged structure are briefly summarized below:

- *Relative incident wave height*: increase of incident wave height H_i and decrease of submergence depth d_r result in a smaller rate of wave transmission and greater wave reflection in case of nonbreaking waves. Larger waves and very shallow water over the structure crest create favourable conditions for wave breaking, leading to significant energy dissipation, and thus smaller wave transmission.
- *Incident wave steepness*: steeper waves in comparison to flatter waves can be forced more easily to break prematurely over the submerged structure, what leads to reduction of the transmitted wave through the energy losses due to breaking. Additionally, concentration of the wave energy in the upper layer of a water column in case of shorter waves (i.e. of small ratio h/L_i) results in a larger transmission in comparison to long waves, for which the energy is uniformly distributed over the water depth.
- *Relative structure width*: width of structure crest B related to incident wavelength L_i is also a parameter controlling wave breaking. Widening of a structure leads to a smaller wave transmission, since energy dissipation due to wave breaking and friction at and through the barrier becomes significant. The minimum structure width results from the width required for wave breaking generation, since its further widening does not enhance the reef

damping performance. Johnson et al. (1951) postulated the following structure width to be optimal for the attenuation of periodic waves:

$$\frac{B}{L_i} = \frac{1}{4}\sqrt{\frac{1-h_r}{h}} \tag{2.16}$$

Dattatri et al. (1978) recommended on the other hand the value of $B/L_i=0.2$-0.3 determined experimentally for regular waves, while Yoshioka et al. (1993) suggested $B/L_i=1.5$ as the optimal value corresponding to the lowest transmission of storm waves over artificial reefs.

- *Structure permeability*: estimation of the influence of the structure porosity on wave transmission is difficult to asses, since the available experiments were performed for structures of differently introduced permeability and the results are very often contradictory. For instance, Dattatri et al. (1978) found the effect of porosity negligible for crushed stone-type submerged structures, unlike Dick and Brebner (1968) who examined barriers made of nested tubes. Powell and Allsop (1985) reported an increase of transmission coefficient by about 20% in case of a barrier of porosity p=0.5 in comparison to its impermeable counterpart (p=0).

The available investigations on the hydraulic performance of various types of low crested structures (shown in Figure 2.5) are briefly summarized in Table 2.2.

c) Analysis of global processes

Theoretical studies

Theoretical models provide a simplified description of the hydraulic performance of submerged structures, including the incident, transmitted and reflected wave field according to linear wave theory. Under the assumption of potential flow, the models do not account for any sources of wave energy dissipation. Since they were developed for shallow-water conditions ($h/L<0.05$), the resultant uniform distribution of horizontal particle velocity over the water column becomes invalid in the vicinity of a submerged barrier. The most important theoretical models to be distinguished are:

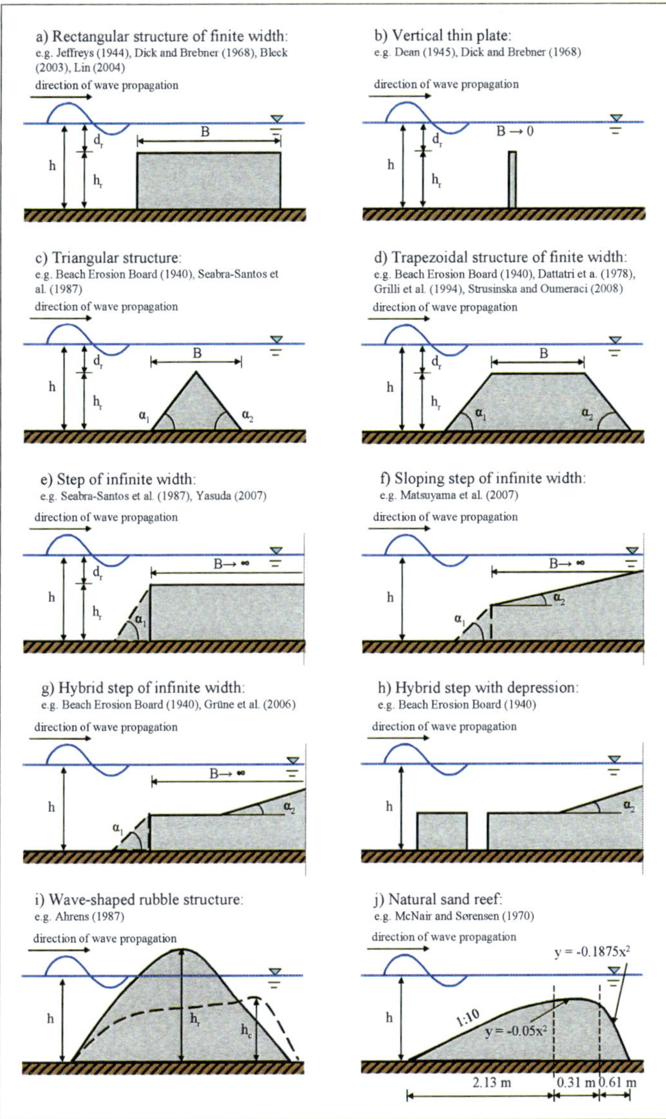

Figure 2.5: Types of low-crested structures studied in the literature

Table 2.2: Selected studies on hydraulic performance of submerged structures

AUTHOR	STUDY TYPE	WAVE TYPE	STRUCTURE TYPE	CONSIDERED PARAMETERS	REMARKS
Lamb (1932)	T, F (LWT)	R, NB	RcS	b, b_1, h, d_r	-
Jeffreys (1944)	T, F (LWT)	R, NB	Im, Rc	H, d_r, B, T_i	formula valid for $h_r/h>0.8$
Dean (1945)	T, F (LWT)	R, NB	Im, TV	d_r	-
Johnson et al. (1951)	T, F (LWT)	R, NB	Im, Rc	h, h_r, L_i	-
	E, D	R, B, NB	Im, Rc	H_i/L_i, B/L_i, h_r/h, h/L_i	optimal structure width and criterion for generation of higher harmonics provided
Nakamura et al. (1966)	E, FM, F	R, B, NB	Im, Rc	d_r/H_i	-
Dick and Brebner (1968)	E, D	R, B, NB	Im (TV, Rc), P	L_i/B	comparison of own results with other existing formulas
Dattatri et al. (1978)	E, D	R, B, NB	Im, P	H_i/L_i, h/L_i, B/L_i, d_r/h, p, slopes	structures of various shapes
Khader and Rai (1980)	E, D	R, NB	Im	H_i/L_i, h/L_i, B/L_i, h_r/h	structures of various shapes
Seelig (1980)	E, F	R, NB	Im, P	d_r, B, H_i, L_i	emerged and submerged structures
Powell and Allsop (1985)	D	I	Im, P	d_r/H_i, p, T_i, B/L_i	emerged and submerged structures, reanalysis of available experimental data
Seabra-Santos et al. (1987)	E, D	S, B, NB	RcS, Trg (Im)	H_i/h, d_r/h	-
Losada et al. (1989)	E, D	S, B, NB	RcS	H_i/d_r	definition of modes of solitary wave behaviour over step
Grilli et al., (1994)	E/N, D	S, B, NB	Im, Tr	H_i/h, h_r/h	emerged and submerged structures
d'Agremond et al. (1996)	F	I	Im, P	d_r, B, H_i, ξ	analysis of available experimental data
Bleck (2003)	E, D, F	R, I, B, NB	Im, Rc	d_r/H_i, B/L_i, H_i/L_i	-
Lin (2004)	N, D (RANS)	S, B, NB	Im, Rc	B/h, h_r/h	wide range of structure widths and heights
Lynett (2007)	N, D (Bq)	S, B, NB	Im, Tr	H_i/h, d_r/h, B/L_i	influence of structure on particle velocity and wave run-up

STUDY TYPE: Bq – Boussinesq-type equations, D – diagrams elaborated, E – laboratory experiments, F – formulas derived, FM – field measurements, LWT – linear wave theory, N – numerical study, RANS – Reynolds-Averaged-Navier-Stokes equations, T – theoretical study; WAVE TYPE: B – breaking waves, I – irregular waves, NB – nonbreaking waves, R – regular waves, S – solitary wave; STRUCTURE TYPE: Im – impermeable structure, P – permeable structure, Rc – rectangular structure, RcS – rectangular step, Tr – trapezoidal structure, Trg – triangular structure, TV – thin vertical plate

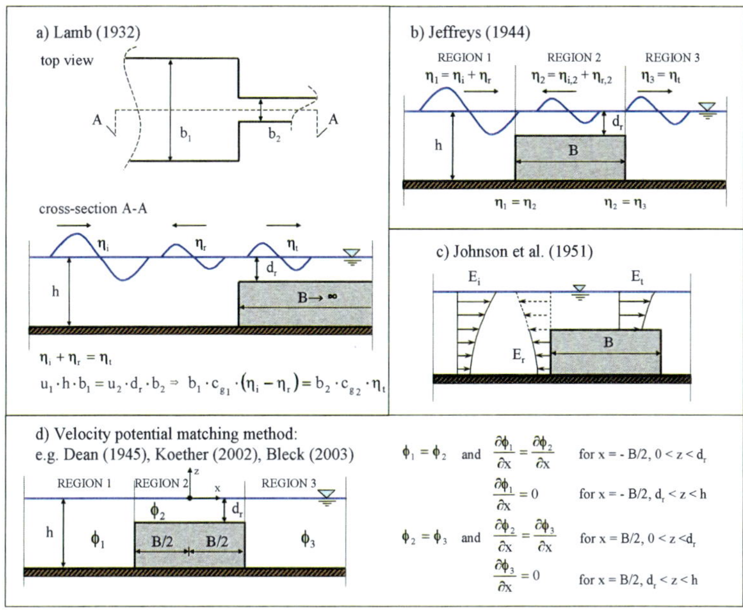

Figure 2.6: Principles and assumption of selected theoretical models for description of global processes induced at submerged obstacles

- Model by Lamb (1932) for propagation of linear waves along a canal of varying width and depth, i.e. from deep water part of the flume of width b_1 and depth h towards a step of width b_2 and depth over its crest d_r (see Figure 2.6a). The wave field in front of the step was assumed to consist of the incident and reflected wave components, while behind this point – of the transmitted wave component.
- Model by Jeffreys (1944) which is a modification of the model by Lamb (1932) accounting for linear wave propagation over a submerged structure of a finite width. The free surface continuity was therefore applied to three sub-domains of wave propagation according to Figure 2.6b: in front of the structure (region 1), over structure crest (region 2) and behind the structure (region 3).
- Model by Johnson et al. (1951) for linear wave propagation over a

submerged structure of a finite width. The hydraulic performance of the barrier is expressed in terms of wave energy: the reflected wave energy is considered as the portion of the incident energy stretched from the bottom to the structure crest, whereas the transmitted wave energy corresponds to the remaining portion over the barrier crest (see Figure 2.6c).

- Models based on velocity potential matching method, developed exemplarily by Dean (1945), Koether (2002) and Bleck (2003). It is required to specify the matching conditions at the boundary among the three sub-domains of wave propagation, which represent the continuity of both pressure and particle velocity, as shown in Figure 2.6d.

The available theoretical formulations for the prediction of the rate of wave transmission and wave reflection are briefly summarized in Table 2.3.

Experimental studies

Most of the available experimental studies were performed in order to estimate transmission characteristic of low-crested structures under different incident wave conditions:

- Impermeable submerged barriers:
    - Regular waves: e.g. Johnson et al. (1951), Nakamura et al. (1966), Dick and Brebner (1968), Dattatri et al. (1978), Khader and Rai (1980), Seelig (1980), Bleck (2003),
    - Irregular waves: e.g. Seabrook and Hall (1998), Daemrich et al. (2001), Bleck (2003),
    - Solitary waves: e.g. Seabra-Santos et al. (1987), Losada et al. (1989), Grilli et al. (1994),
- Permeable submerged barriers:
    - Regular waves: Dick and Brebner (1968), Dattatri et al. (1978),
    - Irregular waves: Powell andAllsop (1985), Ahrens (1987), Van der Meer and d'Angremond (1991), Seabrook and Hall (1998), Daemrich et al. (2001).

Table 2.3: Available analytical formulations for the prediction of wave transmission and wave reflection from impermeable submerged structures

AUTHOR	TRANSMISSION COEFFICIENT K_t [-]	REFLECTION COEFFICIENT K_r [-]	REMARKS
Lamb (1932)	$K_t = \dfrac{2}{1+\dfrac{b_2}{b_1}\cdot\sqrt{\dfrac{d_r}{h}}}$ (2.17)	$K_r = \dfrac{1-\dfrac{b_2}{b_1}\cdot\sqrt{\dfrac{d_r}{h}}}{1+\dfrac{b_2}{b_1}\cdot\sqrt{\dfrac{d_r}{h}}}$ (2.18)	linear waves canal of varying depth and width
Jeffreys (1944)	$K_t = \left(1+0.25\left[\left(\dfrac{h}{d_r}\right)^{1/2}-\left(\dfrac{d_r}{h}\right)^{1/2}\right]^2 \sin^2\left[\dfrac{2B\pi}{T\sqrt{gd_r}}\right]\right)^{-1/2}$ (2.19)	$K_r = \sqrt{\dfrac{0.25\left[\left(\dfrac{h}{d_r}\right)^{1/2}-\left(\dfrac{d_r}{h}\right)^{1/2}\right]^2 \sin^2\left(\dfrac{2B\pi}{T\sqrt{gd_r}}\right)}{1+0.25\left[\left(\dfrac{h}{d_r}\right)^{1/2}-\left(\dfrac{d_r}{h}\right)^{1/2}\right]^2 \sin^2\left(\dfrac{2B\pi}{T\sqrt{gd_r}}\right)}}$ (2.20)	linear waves submerged rectangular structure of finite width
Dean (1945)	$K_t = \dfrac{B_1+B_3}{\sqrt{B_2^2+(B_1+B_3)^2}}$ (2.21)	$K_r = B_2\left[B_2^2+(B_1+B_3)^2\right]^{-1/2}$ (2.22)	linear waves thin, vertical plate
	$B_1 = \displaystyle\int_0^1 \dfrac{e^{\frac{d_r\omega^2}{g}u}}{\sqrt{1-u^2}}du\,,\quad B_2 = \displaystyle\int_0^x \dfrac{\cos\left(\dfrac{\omega^2}{g}u\right)}{\sqrt{u^2+d_r^2}}du\,,\quad B_3 = \displaystyle\int_0^x \dfrac{\sin\left(\dfrac{\omega^2}{g}u\right)}{\sqrt{u^2+d_r^2}}du$	(2.23)	
Johnson et al. (1951)	$K_t = \left(1-\dfrac{B_1(h_r)\cdot\sinh(2kh_r)}{B_2(h)\cdot\sinh(2kh)}\right)^{1/2}$ (2.24)	$K_r = \sqrt{\dfrac{B_1(h_r)\cdot\sinh(2kh_r)}{B_2(h)\cdot\sinh(2kh)}}$ (2.25)	linear waves submerged rectangular structure of finite width
	$B_1(h_r) = \dfrac{1}{2}\left(1+\dfrac{2kh_r}{\sinh(2kh_r)}\right),\quad B_2(h) = \dfrac{1}{2}\left(1+\dfrac{2kh}{\sinh(2kh)}\right)$	(2.26)	

Wave reflection from submerged barriers has attracted very little attention in comparison to the typical coastal protective structures such as breakwaters, seawalls and revetments. The only available studies are those conducted by Ahrens (1987) for irregular waves reflected from a permeable structure and by Bleck (2003) for regular waves reflected from an impermeable barrier.

The most relevant formulas for the prediction of the rate of wave transmission and wave reflection over impermeable submerged structures, developed on the basis of laboratory experiments, are provided in Table 2.4. The details of the studies on wave transmission and reflection in case of permeable counterparts can be found in Strusińska and Oumeraci (2009b).

Table 2.4: Empirically derived transmission and reflection coefficients for impermeable low-crested structures

AUTHOR	FORMULA	REMARKS
TRANSMISSION COEFFICIENT K_t [-]		
Nakamura et al. (1966)	$K_t = 1.0 - 0.8\exp\left(-0.6\frac{d_r}{H_i}\right)$ for $0 < \frac{d_r}{H_i} < 5$ (2.27)	regular waves rectangular structure
Seelig (1980)	$K_t = A_1\left(1-\frac{d_r}{R}\right) - (1-2A_1)\frac{d_r}{R}$ for $0.88 \le B/h \le 3.2$ $R = H_i A_2\left(0.123\frac{L_i}{H_i}\right)^{\left(A_3\sqrt{H_i/h}+A_4\right)}$, $A_1 = 0.51 - \frac{0.11B}{h_r}$ A_2=1.997, A_3=0.498, A_4=-0.185 (2.28)	regular waves rectangular structure
Losada et al. (1989)	$K_t = H_{min}/H_i = -0.17\,H_i/d_r - 1.04$ (2.29)	solitary wave infinite step
d'Angremond et al. (1996)	$K_t = -0.4\frac{d_r}{H_i} + \left(\frac{B}{H_i}\right)^{-0.31}\left[1-\exp(-0.5\xi)\right]0.8$ (2.30) $\xi = \tan\alpha/\sqrt{H_i/L_i}$	irregular waves analysis of available experimental data
Bleck (2003)	$K_t = 1 - 0.83\exp(-0.72\,d_r/H_i)$ (2.31)	regular waves rectangular structure
REFLECTION COEFFICIENT K_r [-]		
Bleck (2003)	$K_r = 0.57\exp(-0.23\,d_r/H_i)$ (2.32)	regular waves rectangular structure

Numerical studies

At the early time of computational modelling, simple and not time-consuming depth-averaged long wave equations were employed to investigate wave-structure interaction, mostly under nonbreaking conditions (e.g. Goring, 1979; Seabra-Santos et al., 1987). As the numerical modelling advanced, more accurate and complex theories, such as Boussinesq-type equations and Reynolds-Averaged-Navier-Stokes model, were used to improve the description of the global and local processes, capable of reproducing wave breaking and vortex generation (e.g. Chang et al., 2001; Liu and Cheng, 2001; Lin, 2004; Lynett, 2007). The types of numerical models for water wave propagation currently used, are provided in Section 2.4.

Of the utmost importance are however the studies that indicate the application of available wave theories to the nonlinear wave transformation over low-crested structures. For instance Goring (1979) reproduced own laboratory experiments on solitary and cnodial wave transmission over a submerged step by means of wave theories of different nonlinear and frequency dispersion properties: linear dispersive, linear nondispersive, nonlinear dispersive (standard Boussinesq model) and nonlinear nondispersive (nonlinear shallow water NSW equations). He found that the nonlinear dispersive wave model provides the best prediction of the transmitted wave profile, since wave nonlinearity dominates at the abrupt change of water depth, while the frequency dispersion effects become relevant with the propagation distance over the barrier.

Further results of solitary wave scattering at a horizontal step were reported by Goring and Raichlen (1992), which once more confirmed the advantage of the Boussinesq model, accounting for both nonlinear and dispersion effects, over the other water wave theories.

Ohyama et al. (1995) compared the experimental data on the transmission of regular waves over a submerged trapezoidal structure, accompanied by the generation of higher harmonics with numerical predictions by means of fully nonlinear potential model, Stokes 2nd order theory, standard Boussinesq and Boussinesq-type model by Nwogu (1993). For short small-amplitude waves, both the nonlinear potential model and the Stokes theory predicted wave behaviour well. With the increase of wave nonlinearity and decrease of kh, the Stokes theory tended to overestimate the height of higher harmonics and underestimate the number of the generated higher order components. A similar tendency was also observed for the original Boussinesq equations. The best agreement with the laboratory results was achieved by means of the Nwogu's (1993) Boussinesq theory, characterized by the improved dispersion characteristics that allowed to predict correctly the transfer of the energy into higher frequencies.

Gobbi and Kirby (1999) reproduced the experiments by Beji and Battjes (1993) on the propagation of periodic waves over a submerged trapezoidal-shaped bar. The comparison between the fully nonlinear, extended Boussinesq model by Wei et al. (1995) of accuracy $0(\mu^2)$ and the high order Boussinesq model by Gobbi and Kirby (1999) of accuracy $0(\mu^4)$ indicated the influence of both nonlinear and dispersion effects on the transmitted wave characteristics.

2.3.3 Characteristics of local processes

Among the local processes, two phenomena are of the utmost importance for the wave damping performance of a low-crested structure: wave breaking and energy transfer into higher frequencies, termed hereafter the nonlinear effects.

Unlike waves that normally break at a shore, waves approaching a submerged barrier are forced prematurely to break in the reduced water depth over the obstacle crest. Reduction of the incident wave height results from the energy dissipation due to turbulence and energy conversion into heat. Description of the breaking conditions over a low-crested structure requires specification of both wave breaking criteria and wave breaker types.

Nonlinear effects are induced as a wave propagates from a deep portion of water into a region of reduced water depth, what is the case of both submerged barriers of an infinite width (e.g. steps) and of a finite width (e.g. artificial reefs). The short-period waves, generated by energy transfer into higher frequencies over a submerged structure, can be more easily attenuated by the successive protective lines of the multi-line defence system against tsunami, mentioned in the introductory chapter.

a) Wave breaking and breaker types

Wave breaking criteria

Wave breaking criteria can be generally classified into two main groups:

- *Wave steepness criteria* valid for wave breaking in deep water, controlled by deep-water wave steepness:

$$\gamma_d = (H_o/L_o)_b \tag{2.33}$$

- *Water depth limit criteria* (or *breaker depth indices*) applicable to wave breaking in shallow water, controlled by the water depth at breaking:

$$\gamma_s = (H_s/h)_b \tag{2.34}$$

where subscript "o" denotes the deep-water wave conditions, subscript "s" the shallow-water wave characteristics and subscript "b" the conditions at breaking.

The most relevant criteria for wave breaking over a horizontal bottom are provided in Table 2.5. The influence of the plane sloping sea bed on wave breaking is incorporated into the breaking limits either through bottom slope tanα (e.g. Galvin, 1969), or through wave steepness/Irribarren number (e.g. Singamsetti and Wind, 1980), or through a combination of these two parameters (e.g. Weggel, 1972). More details can be found in Strusińska and Oumeraci (2009a).

Table 2.5: Wave breaking criteria on horizontal bottom

AUTHOR	BREAKING CRITERION	REMARKS
WAVE STEEPNESS CRITERIA $\gamma_d=(H_o/L_o)_b$ (DEEP WATER)		
Michell (1893)	$\gamma_d = 0.142 = 1/7$ (2.35)	progressive wave in deep water theoretical
Goda (1970)	$\gamma_d = A_1\left\{1-\exp\left[-1.5\frac{\pi h_b}{L_o}\left(1+15\tan\alpha^{4/3}\right)\right]\right\}$ (2.36) with A_1=0.17 for regular waves	regular wave sloping sea bottom $0.02 \le \tan\alpha \le 0.1$ experimental
WATER DEPTH LIMIT CRITERIA $\gamma_s=(H_s/h)_b$ (SHALLOW WATER)		
McCowan (1891)	$\gamma_s = 0.78$ (2.37)	solitary wave theoretical
Miche (1944)	$\gamma_s = 0.88$ (2.38) from: $H_{max} = 0.142L\tanh(kh)$	regular waves theoretical
Lenau (1966)	$\gamma_s = 0.83$ (2.39)	solitary wave theoretical
Le Mehaute et al. (1968), Nelson (1987)	$\gamma_s = 0.55$ (2.40)	regular waves experimental

Criteria for wave breaking over different types of low-crested structure have been predominantly developed for regular/irregular wave conditions (e.g. Smith and Kraus, 1991; Iwata et al., 1996; Bleck, 2003). Breaking criteria for a solitary wave can be represented by those determined by Hara et al. (1992) on a basis of numerical investigation. The summary of the breaking limits at submerged structures is provided in Table 2.6.

Table 2.6: Criteria for wave breaking over submerged structures

AUTHOR	BREAKING CRITERION	REMARKS
Smith and Kraus (1991)	$\gamma_s = H_b/h_b = 0.41 + 0.98\xi_o$ for $0.3 \le \xi_o \le 0.85$ $\gamma_s = H_b/h_b = 1.45 - 0.22\xi_o$ for $1.6 \le \xi_o \le 3.5$ $\xi_r = \tan\alpha_1 / \sqrt{H_o/L_o}$ (2.41)	regular waves triangular bar experimental
Iwata et al. (1996)	a) for h/L_i=0.13 and for $0.1 \le B/L_i \le 0.6$: $\frac{H_b}{d_r} = 0.3656 + 0.3668 e^{\left(\frac{-10B}{L_i}\right)}$ (2.42) b) for h/L_i=0.2 and for $0.1 \le B/L_i \le 0.6$: $\frac{H_b}{d_r} = 0.3641 + 0.2313 e^{\left(\frac{-10B}{L_i}\right)}$ (2.43) c) for h/L_i=0.4 and for $0.1 \le B/L_i \le 1.0$: $\frac{H_b}{d_r} = 0.4417 + 0.2252 e^{\left(\frac{-10B}{L_i}\right)}$ (2.44)	regular waves rectangular structure with finite width experimental
Bleck (2003)	a) for regular waves (modified Miche's, 1944 criterion) $\left(\frac{H_i}{L_i}\right)_b = 0.142 \tanh\left(0.74 \frac{2\pi}{L_i} d_r\right)$ (2.45)	regular and irregular waves rectangular structure of finite width experimental
Hara et al. (1992)	a) critical height for infinite step for $0.2 \le h_r/h \le 0.6$: $\frac{H_{crit}}{h} = 1.012 - 1.063\left(\frac{h_r}{h}\right)^{0.46}$ (2.46) b) critical height for barrier of finite width B for $0.2 \le \gamma \le 1.4$: $\frac{H_{crit}}{h} = 0.952 - 0.591\gamma^{0.76}$ $\gamma = \left(\frac{B}{h} + \frac{h_r}{2h}\tan\alpha_1\right)^{0.4} \frac{h_r}{h}$ (2.47)	solitary wave infinite step trapezoidal/ rectangular step of finite width numerical
Hara et al. (1992)	a) breaker depth index for infinite step: $\frac{H_b}{h_b} = \left(5.885 - 5.09\xi_s^{*0.133}\right)\frac{h}{h - h_r}$ $\xi_s^* = \frac{(h_r/h)^{0.1}}{(H/h)^{0.4}}$ (2.48) b) breaker depth index for trapezoidal step of finite width B for $0.2 \le \xi_t^* \le 2.0$: $\frac{H_b}{h_b} = 1.039 - 0.463\xi_t^{*0.133}$ $\xi_t^* = \left[\frac{B}{h} + \frac{(h_r/h)}{3.5}\tan\alpha_1\right]^{0.2} \frac{(h_r/h)}{(H/h)^{0.4}}$ (2.49) c) breaker depth index for rectangular step of finite width B: replace ξ_t* by $\xi_r^* = \frac{(B/h)^{0.2}(h_r/h)}{(H/h)^{0.4}}$ (2.50)	solitary wave infinite step trapezoidal/ rectangular step of finite width numerical

Types of wave breakers

The very first classification of breaker types was elaborated for waves breaking at natural beaches by Patrick and Wiegel (1954), and further modified by Galvin (1968) by introducing fourth breaker termed collapsing breaker:

- *Spilling breaker* typical for very gently sloping or flat beaches and steep waves. Wave crests break gradually over a large distance in a manner resembling a frothy bore. The white water spreads from the wave crest to the down parts of the water front.
- *Plunging breaker* typical for either mildly steepening beaches with waves of small wave steepness or steep beaches with steep waves. Wave crests curl over and plunge into the water in front like a jet, causing a dramatic splash advancing like a bore. Large quantity of air is trapped between the jet and the wave front, producing spray and white water.
- *Surging breaker* common for very steep beaches and waves, whose height is small in comparison to the wavelength. Only wave base breaks gently and undergoes significant reflection, since the lower parts of the wave surge faster than the jet can be formed.
- *Collapsing breaker* combining the features of the plunging and surging breakers. Wave crest does not break, yet the wave base, creating a foamy surface, advancing up to the beach. The bottom morphology and wave steepness are identical to the case of the surging breaker.

Since breaking conditions over low-crested structures differ from those over natural beaches, a new classification of wave breakers is required. The type of breaking of regular waves at submerged structures was observed to be strongly influenced by the interaction of reverse flow, induced by a wave already passed over the barrier, and a new incoming wave in the wave train (Bleck, 2003):

- *Spilling breaker* reducing wave energy by about 29%. Wave breaks over the barrier and generates a bore. Little turbulence over the structure is produced and the reverse flow is also weak. The relative wave height H_i/d_r is small in this case. This form of wave breaking is similar to the spilling breaker observed over a sloping beach.
- *Two-step breaker* or *double breaker* dissipating up to 42% of the incident

energy. The zone of breaking and turbulence develop first at the seaward edge of the barrier when the reverse flow interacts with the trough of the successive wave. This turbulent zone remains still at the same place (however becomes weaker and weaker) as the wave enters the shallow water over the reef and breaks eventually. The wave propagates then downstream as a turbulent bore, followed by an increased water level. The return flow influences the incoming wave by increasing the ratio of the wave height and the wavelength.

- *Drop-type breaker* with the decrease of the energy by approximately 46%. The return flow is strong enough to separate at the seaward edge of the reef, generating a vortex, and finally leading to wave breaking before the wave approaches the obstacle. This phenomenon resembles overtopping of a dam. Since waves propagating over the reef resemble a bore, additional turbulent effects appear in a manner similar to the spilling breaker.

Breaker types of a solitary wave propagating over a low-crested structure resembles generally those observed at natural beaches and are classified as follows (Losada et al., 1989; Grilli et al., 1994):

- *Spilling breaker* of the same properties as the breaker observed over natural beaches.
- *Transition breaker* representing a transition form between spilling and plunging breakers.
- *Plunging breaker* of the same properties as the breaker observed over natural beaches.

Cooker et al. (1990) and Grilli et al. (1994) distinguished apart from the *forward breaking* of a transmitted solitary wave (taking either the spilling or the plunging form), *backward breaking* of the wave tail, i.e. in the direction opposite to the propagation of the transmitted wave, induced once the main wave passed the structure.

Characteristic of breaker types (surf similarity parameter)

Properties of wave breakers can be quantitatively described by using the ratio of sea bottom slope to wave steepness, termed generally the *surf similarity parameter*:

$$\xi_o = \frac{\tan\alpha}{\sqrt{H_o/L_o}} \tag{2.51}$$

Such a parameter was first determined by Irribarren and Nogales (1949) for breakers developed over natural beaches and commonly termed the *Irribarren number*.

However, the determination of the breaking properties on the basis of offshore wave characteristics was found to be incorrect. Therefore, successive modifications of the original Irribarren number were introduced to account for the local wave characteristic, e.g. Galvin (1968), Battjes (1974), Yoo (1986), Schüttrumpf (2001) (see Table 2.7).

Similar parameters were also developed for the case of wave breaking over low-crested barriers, by replacing the sea bottom slope by the obstacle geometry. Smith and Kraus (1991) determined the breaker number for irregular wave conditions as a function of the seaward slope of the barrier α_1:

$$\xi_r = \frac{\tan\alpha_1}{\sqrt{H_o/L_o}} \tag{2.56}$$

with the corresponding breaker types: spilling breaker for $\xi_r<0.4$, plunging breaker for $0.4<\xi_r<1.2$ and surging breaker for $\xi_r>1.2$. In comparison to the breakers observed at a natural beach, there is a shifting of the boundaries among the breakers – incident waves that would normally break as a spilling breaker on a gently sloping beach, break as a plunging breaker in the case of a barred sea bottom. Therefore, a beach with underwater features tends to act in a similar manner to a beach of a very steep slope.

Hara et al. (1992) on the other hand developed three different breaking numbers for solitary waves, depending on the bottom bathymetry:

Plane slope: $$\xi^* = \frac{\tan\alpha}{(H_i/h)^{0.4}} \tag{2.57}$$

Submerged step: $$\xi_s^* = \frac{(h_r/h)^{0.1}}{(H_i/h)^{0.4}} \tag{2.58}$$

Trapezoidal barrier: $$\xi_t^* = \left[\frac{B}{h} + \frac{(h_r/h)}{3.5}\tan\alpha_1\right]^{0.2} \frac{(h_r/h)}{(H_i/h)^{0.4}} \qquad (2.59)$$

Rectangular barrier: $$\xi_r^* = \frac{(B/h)^{0.2}(h_r/h)}{(H_i/h)^{0.4}} \qquad (2.60)$$

Table 2.7: Surf similarity parameter for waves over sloping beaches and corresponding breakers

AUTHOR	DEFINITION	BOTTOM SLOPE	TYPE OF WAVE BREAKER			
			SPILL.	PLUNG.	COLL.	SURG.
Battjes * (1974)	$\xi_o = \frac{\tan\alpha}{\sqrt{H_o/L_o}}$ (2.52)	1:5 1:10 1:20	$\xi_o \le 0.5$	$0.5<\xi_o<3.3$	-	$\xi_o \ge 3.3$
Battjes ** (1974)	$\xi' = \frac{\tan\alpha}{\sqrt{H_b/L_o}}$ (2.53)	1:5 1:10 1:20	$\xi' \le 0.4$	$0.4<\xi'<2.0$	-	$\xi' \ge 2.0$
Yoo (1986)	$\beta = \frac{\xi_b^2}{\pi kh}$ (2.54) $\xi_b = \frac{\tan\alpha}{\sqrt{H_b/L_b}}$	-	$\beta \le 0.2$	$0.2<\beta<2.1$	-	$\beta \ge 2.1$
Schüttrumpf (2001)	$\xi_s = \frac{\tan\alpha}{\sqrt{H_{local}/L_o}}$ (2.55)	1:6	-	$\xi_s \le 2.1$	$2.1<\xi_s<2.8$	$\xi_s \ge 2.8$
		1:4	-	$\xi_s \le 2.4$	$2.4<\xi_s<3.1$	$\xi_s \ge 3.1$
		1:3	-	$\xi_s \le 2.6$	$2.6<\xi_s<3.3$	$\xi_s \ge 3.3$

* Re-analysis of Galvin's (1968) results in terms of Irribarren and Nogales' (1949) surf similarity parameter ξ_o
** Re-analysis of Galvin's (1968) results in terms of Battjes' (1974) surf similarity parameter ξ'
SPILL. – spilling, PLUNG. – plunging, COLL. – collapsing, SURG. – surging breakers

b) Characteristic of nonlinear processes

Higher harmonics generation in case of regular and irregular waves

Due to nonlinear interaction of incident regular/irregular waves with a low-crested structure, generation of multiple wave crests can be observed as wave energy is transferred into higher frequencies. Over the barrier crest, these waves termed *bound harmonics* are phase-locked with the fundamental waves, since both propagate at the same speed. In case of a submerged structure of a finite width, the bound harmonics continue propagating in the deeper water behind the obstacle as *free higher harmonic waves* of second and higher order.

The results of the experimental and numerical studies on the wave energy transfer into higher frequencies, occurring over submerged structures of finite widths, can be summarized as follows:

- Amplitudes of the higher harmonics increase in the reduced water depth over the structure crest and tend to be constant as they are transmitted into the deeper water behind the barrier of a finite width (Goda et al., 2000).
- In the deeper water behind the barrier, wave nonlinearity becomes very weak so that the bound harmonics cannot exist and are transmitted further as free higher harmonics.
- Fourier Transformation applied to the laboratory data by Bleck (2003) indicated broadening of the transmitted wave spectrum in comparison to the incident wave spectrum. The frequency of the incident and fundamental transmitted waves were the same, however the height of the primary fundamental transmitted wave was reduced (see Figure 2.3). The frequency of the second harmonics corresponded to the double fundamental frequency.
- Increase of both structure width, wave period and wave amplitude result in a stronger energy transfer into higher frequency, particularly in case of the 5th and 6th order wave components, as shown by Ting et al. (2005). This analysis, unlike most of the available studies, included harmonics up to the 6th order.

Generation of higher harmonics was also reported in case of submerged structures of infinite widths, represented by underwater steps (e.g. Isobe et al., 1996; Goda et al., 2000). The observed spatial variation of the amplitude of second and higher order harmonics above the step is related to their beat length (the recurrence distance). According to Ohyama and Nadaoka (1994), the beat length of the 2nd order harmonics can be determined as follows:

$$\lambda_2 = 2\pi/(k_2 - 2k_1) \tag{2.61}$$

where k_1 is the wave number of the bound waves, and k_2 the wave number of the free waves in the second harmonics, respectively. The maximum amplitude of the second harmonics was observed for the shelf width approximately half of the beat length. The rate of the energy transfer into higher frequencies depends on the ratio of the barrier width to the beat length of the “i-th” harmonic components B/λ_i considered.

Solitary wave fission

A similar phenomenon to the generation of higher harmonics is reported when a single solitary wave disintegrates above a submerged barrier into a train of solitary waves of decreasing heights (*solitons*), trailed by oscillatory waves. This phenomenon, commonly termed *solitary wave fission*, has been however investigated primarily for the case of submerged barriers of infinite width, such as steps (e.g. Madsen and Mei, 1969; Losada et al., 1989). It was explained theoretically (e.g. Tappert and Zabusky, 1971; Johnson, 1972) and reproduced by numerical studies based predominantly on depth-averaged long wave equations, in which the effect of soliton breaking was not incorporated (e.g. Madsen and Mei, 1969; Seabra-Santos et al., 1987).

The only study available, which considers the effect of a structure width on the wave fission, is that by Seabra-Santos et al. (1987). Unlike the infinite step for which at least two solitons were recorded, regular solitary wave transmission was observed in the case of a triangular submerged structure. The number and amplitudes of the solitons emerged above the step under laboratory conditions were found to be in a very good agreement with the results obtained by means of the theoretical formula by Germain (1984) and Kabbaj (1985), applied to both nonbreaking and breaking waves (see Eq. 2.64).

Losada et al. (1989) observed the following modes of solitary wave behaviour when passing over an infinite step:

- *Wave propagation with insignificant change of shape* (weak distortion) observed for weakly nonlinear waves that maintained approximately a constant shape when propagating over the barrier.
- *Fission of the leading transmitted wave* observed for weakly nonlinear dispersive waves, which disintegrated into solitons over the barrier without breaking of the leading wave.
- *Wave fission followed by the breaking of the leading transmitted wave as a spilling breaker*, typical for strongly nonlinear dispersive waves, which disintegrated into solitons before the leading wave started breaking.
- *Breaking of the leading wave as a plunging breaker without wave fission*, observed for strongly nonlinear dispersive waves, which propagated as an undular bore consisting of several waves with the leading edge still breaking

(resembling rather a turbulent undular bore).

Liu and Cheng (2001) extended the analysis of the solitary wave fission over an infinite step to breaking wave conditions, performed using the Reynolds-Averaged-Navier Stokes (RANS) equations. Additionally, nonbreaking wave conditions in the experiment of Seabra-Santos et al. (1987) were reproduced, with a very good agreement with the data. Two cases of breaking conditions were considered:

- *Fission of the incident wave and breaking of the leading fission over the step*: the amplitude of the leading soliton was reduced significantly due to breaking and reached a constant value as the breaking process was completed. The amplitude of the second emerged soliton remained constant, unlike the amplitude of the third wave which tended to increase with the travel distance. In comparison to the nonbreaking case, the height of the leading solitons was smaller, unlike the heights of the second solitons that were not affected by the breaking event and were of the same order.
- *Breaking of the incident wave in front of the step*, characterized by propagation of the broken wave in a form of an undular bore. Unlike the study of Losada et al. (1989), wave fission was generated after the breaking event started. The amplitudes of the first and second solitons were evidently smaller than those observed during the breaking event induced over the step.

Solitary wave fission was also generated over submerged structures of finite widths in numerical investigations of Lin (2004), focused on the determination of the rate of wave transmission and reflection for rectangular obstacles of different configurations. It was briefly mentioned that in case of both nonbreaking and breaking waves, only the first soliton maintained the characteristic of a solitary wave, while the further solitons propagated as oscillatory waves once entering the deeper water behind the structure.

The importance of the fission phenomenon during tsunami wave propagation from deep water towards a shore over a continental shelf was indicated by laboratory experiments of Matsuyama et al. (2007) and Yasuda (2007). Despite the fact that the tsunami was generated as a long sinusoidal wave of one cycle only, it disintegrated into a number of short period waves in the same manner as a solitary wave does. Matsuyama et al. (2007) reported that the wave disintegration point moved forward from the wave peak to the wave front as the wave period increased.

Several theoretical formulations are available that can be applied to predict number of solitons N emerged over a submerged step of an infinite width as a function of water depth over the step (d_r) and in front of the step (h):

- Tappert and Zabusky (1971):

$$d_r / h = \left[0.5N(N-1)\right]^{-4/9} \tag{2.62}$$

- Johnson (1972):

$$d_r / h = \left[0.5N(N+1)\right]^{-4/9} \tag{2.63}$$

  If the number of the solitons obtained by Eq. (2.63) lies between two integers N_o and N_o+1, N_o+1 solitons will emerge, trailed by oscillatory waves. If the number of solitons is already an integer, only solitons, without the oscillatory wave train will be generated.

- Germain (1984) and Kabbaj (1985):

$$M = 0.5\left\{\left[1+\frac{16(h/d_r)^{5/2}}{1+(h/d_r)^{1/2}}\right]^{1/2} - 1\right\} \tag{2.64}$$

  $N = |M|$, N is the largest integer $\leq M$.

2.3.4 Implications for this study

Based on the results of the detailed study on the hydraulic performance of low-crested structures presented in Strusińska and Oumeraci (2009b) and the brief overview of this problem provided in this section, the following conclusions can be made:

- The full characteristics of the hydraulic performance of low-crested structures are available only for regular and irregular waves. In case of a solitary wave, which can approximately represent a tsunami wave, only selective aspects of wave-structure interaction have been yet investigated.
- A strong nonlinear interaction between a tsunami-like solitary wave and an artificial reef would be expected, including wave breaking generation and wave disintegration into solitons. Both nonlinear and frequency dispersion

effects should be considered in this case as indicated by the available experimental and numerical investigations.

- The available analytical potential flow models are not applicable to estimate the damping performance of a submerged structure, since wave energy dissipation is not included. Additionally, due to the complexity of the phenomena generated at the structure, a numerical model capable of dealing with the local and global processes would be required in this study to determine the hydraulic performance of the structure.
- The advantage of the fully nonlinear, extended Boussinesq-type models over the other wave theories, particularly the nonlinear shallow-water (NSWE) equations was indicated by the available studies. The weakness of the NSWE model lies in the following: (i) poor prediction of the incipient wave breaking and associated energy dissipation, (ii) incorporation of nonlinear effects only what makes it incapable of dealing with the process of wave disintegration over a submerged structure, (iii) no representation of the vertical velocity field which is assumed uniform over water depth. The latter assumption becomes invalid at the vicinity of a submerged structure.
- In view of the relevant research gaps, performance of own laboratory experiments would be required to provide a reliable criterion for solitary wave breaking over a submerged structure of a finite width as well as to understand the process of wave fission occurring for this structure type.
- The hydraulic functioning of the reef should be determined by the rate of wave transmission, wave reflection and wave energy dissipation, expressed in terms of wave energy (since definition of an adequate transmitted wave height in case of solitary wave fission is difficult).
- Both wave breaking and vortex generation contribute to the significant dissipation of the incident wave energy. The other source of energy losses, resulting exemplarily from bottom friction, can be neglected.

2.4 Available numerical models for water wave propagation

The main features of the available numerical models for water wave propagation will be presented in this section. Since the detailed characteristics of the water wave theories, representing the core of the programmes, were already provided in

Section 2.2, the numerical schemes, the required computational effort and the fields of applications of the models will primarily be considered in this section.

Classification of the numerical models encompasses generally two main groups: *phase-averaged* and *phase-resolving* models (Liu and Losada, 2002). The phase-averaged models predict wave characteristic for one full wave phase according to the spectral energy balance equation, and since they require relatively low spatial resolution, they can be applied to large-scale coastal problems. In the phase-resolving models, the properties of wave motion can be determined at any point along a wave from the time-dependent, depth-integrated mass and momentum conservation equations. The spatial and temporal resolution is in this case finer (about 10-100 time steps for each wave period) what results in the practical applications restricted to a computational domain in the range of 1-10 km.

2.4.1 Direct Numerical Simulation (DNS)

All physical processes involved in water wave propagation and nearshore transformation can be accurately modelled only by means of the Navier-Stokes equations, solved numerically by the technique termed the Direct Numerical Simulation (DNS). As already mentioned, the computational effort involved in the very high spatial and time resolution of turbulent effects and energy dissipation due to breaking, structure-induced turbulence, etc., makes the models inapplicable for simulations at a larger scale even in the very near future (Liu and Losada, 2002).

2.4.2 RANS and LES models

a) RANS model

The numerical technique employed for solving the classical Reynolds equations is termed the *Reynolds-Averaged-Navier-Stokes* (RANS) method. Due to the averaging of the Navier-Stokes equations in time, incorporation of an appropriate closure is required for computing the Reynolds stresses, accounting for the unresolved small-scale turbulent effects. These closure models can be introduced into the equations as the Prandtl mixing length, eddy viscosity model or more sophisticated models such as k-ε (consisting of the transport equations for turbulent kinetic energy k and turbulent dissipation rate ε).

The numerical code COBRAS (Cornell Breaking Wave and Structures) represents one of the most advanced RANS models. It was developed by Liu and Lin (1997) at the Cornell University (USA) from the model RIPPLE, originally created for the NASA, and further extended to model breaking waves and flow through porous media. The RANS technique is coupled with the Volume of Fluid (VOF) method in order to track free surface elevation and the k-ε model governing the Reynolds stresses. Apart from the free surface displacement, a detailed prediction of the pressure and particle velocities field, vorticity, turbulent viscosity, kinetic energy and turbulent dissipation is provided.

b) LES model

The Large Eddies Simulation (LES) technique was originally introduced by Smagorinsky to model atmospheric air flow on the basis of the Navier-Stokes equations averaged in space. The resultant governing equations are resolved explicitly for large eddies only, while smaller eddies are approximated by means of a sub-grid scale model. This can be correct when assuming that large eddies are dependent on the flow geometry in contrast to small eddies which have universal properties, according to the Kolmogorov's (1942) theory on the eddies self similarity. The Smagorinsky sub-grid model is commonly used in practice and it requires incorporation of eddy viscosity term into the equations to account for the unresolved turbulent effects. Unlike the RANS that models time-averaged flow characteristics, the LES method provides instantaneous flow properties at much higher computational effort.

2.4.3 Potential flow models

Fully nonlinear potential flow models allow to predict the evolution of highly nonlinear waves up to breaking conditions. The accurate modelling of breaking wave together with the associated wave energy dissipation is out of the capabilities of the models, since the flow is assumed to be irrotational. The boundary element method (BEM) was originally applied by Longuet-Higgins and Cokelet (1976) to solve the governing equations, and further improved at the University of Rhode Island (USA) by Grilli et al. (1989). This approach is based on the boundary integral equation method (BIEM) used to compute the governing equations, that are integrated in time by means of an explicit Lagrangian time marching. Unlike the depth-integrated equations, the free surface displacement is not represented by a single value, and therefore the shape of the plunging jet during breaking event can be accurately predicted. However, the model breaks down once the plunging jet touches the frontal water free surface. The model has been successfully applied to both 2D and 3D wave propagation (e.g. Grilli et al., 1989; Grilli et al., 1994).

2.4.4 Models based on nonlinear shallow-water wave equations

The available numerical models based on the nonlinear shallow water wave theory have recently been incorporated as an integral part of the tsunami early warning systems for real-time tsunami forecasting as well as preparation of both tsunami inundation and evacuations maps. The information on the tsunami arrival time, height and extension of the inundated area can be provided immediately after detection of a tsunami event due to the low-computational costs at which the simulations are performed. Such models are exemplarily represented by:

- COMCOT (Cornell Multi-grid Coupled Tsunami Model) developed by Liu et al. (1994) at the Cornell University (USA),
- MOST (Methods of Splitting Tsunami) developed by Titov and Synolakis (1998) at the University of Southern California (USA),
- TSUNAMI-N2 developed by Shuto and Imamura at the Tohoku University (Japan).

As already mentioned in Section 2.2, the fundamental weakness of these models lies in the prediction of breaking waves that is treated purely numerically by means of the Lax-Wendroff scheme: (i) the frozen wave front, representing breaking wave, covers too small region in comparison to the breaking observed in nature, (ii) the breaking criterion is determined as the position of the frozen front in respect to numerical offshore boundary, (iii) due to the purely nonlinear character of the equations, the incipient breaking might be predicted incorrectly over the horizontal bottom, since wave height evolves continuously, (iv) the rate of energy dissipation is controlled by means of numerical coefficients.

2.4.5 Models based on Boussinesq-type equations

The two following Boussinesq-type numerical models are intensively employed for a wide range of coastal application aspects:

- Model FUNWAVE developed by Kirby et al. (1998), based on the fully nonlinear, extended Boussinesq-type equations by Wei et al. (1995), with the horizontal particle velocity computed at arbitrary water level, postulated by Nwogu (1993) to be z_α=-0.531h.
- Model COULWAVE developed by Lynett (2002), based on the fully nonlinear, extended Boussinesq-type equations by Liu (1994), with Nwogu's (1993) concept of the horizontal velocity evaluation at water level z_α=-0.531h. The governing equations were further modified to account for modelling of waves generated by submarine landslides (Lynett, 2002).

Both models were established independently and are freely available. Their common features can be summarized as follows:

- Computed parameters: free surface displacement and horizontal velocity field.
- Modelling of energy dissipation due to wave breaking and bottom friction by introducing additional dissipative terms in the momentum balance equation (for details and differences in the specified parameters see Kennedy et al., 2000 and Lynett, 2002):
  - wave breaking criterion controlled by the critical value of the vertical speed of free surface elevation,
  - energy dissipation modelled by means of the modified eddy viscosity concept of Zelt (1991),
  - frictional losses computed according to the quadratic friction law.
- Finite difference method for discretization of the governing equations in space and 4th order Adams-Bashforth-Moulton corrector-predictor scheme for time stepping (consisting of the 3rd order Adams-Bashforth predictor step and 4th order Adams-Moulton corrector step). The method of grouping of the velocity time derivatives in the momentum conservation equations introduced in COULWAVE is believed to improve stability of the model (Lynett and Liu, 2002) in comparison to that used by Wei et al. (1995).

In contrast to the FUNWAVE model, an additional option of wave generation due to underwater landslide is incorporated in COULWAVE. Moreover, improvement of the velocity field up to kh~6 was achieved by introducing the multi-layer approach. Recently, high-order finite volume method using the shock-capturing Reimann solver for the leading order flux terms was implemented to improve the model stability (Kim et al., 2009).

2.4.6 Hybrid models

The future of the modelling of hydrodynamical coastal processes is however conditioned by the development of the so-called *hybrid models* that couple two models of different properties: a less accurate model for large scale application in the farfield and a very accurate flow model applied in the region of interest, where detailed wave motion and flow characteristics are required. This technique allows to model large-scale wave propagation with a significant reduction of the computational costs, with no losses of the solution accuracy at the same time. The available hybrid models are for instance:

- Model COULWAVE by Lynett (2002) based on the fully nonlinear, extended Boussinesq equations (Liu, 1994) coupled with COBRAS by Liu and Lin (1997), based on the RANS-VOF approach (see e.g. Sitanggang et al., 2006).
- Fully nonlinear potential model by Grilli et al. (1989), based on the BIM method coupled with Navier-Stokes equations (see e.g. Grilli et al., 2004).

2.4.7 Implications for this study

In view of the required computational effort, the depth-averaged models have the advantage over the more detailed flow models (such as DNS, RANS or LES) to be capable of dealing with large-scale coastal problems. Since the hybrid models are still at the development stage, indeed the numerical codes based on the nonlinear shallow water equations and Boussinesq-type theories are considered as the most appropriate tools to predict nearshore tsunami evolution. However, in this study, a strong nonlinear interaction of a long tsunami wave with a submerged structure is expected, including wave breaking over the reef and generation of shorter wave components due to nonlinear effects. Therefore, a numerical model that accounts for both nonlinear and dispersive effects would be preferable, with a realistic prediction of wave breaking event and the associated energy dissipation.

Fully nonlinear, extended Boussinesq equations by Liu (1994) and Wei et al. (1995), which are capable of treating both highly nonlinear and dispersive waves, were implemented in COULWAVE by Lynett (2002) and FUNWAVE by Kirby et al. (1998). Despite very similar features between these two numerical codes, COULWAVE was selected to perform the numerical analysis of the hydraulic performance of artificial reef under tsunami attack, since:

- It has been widely employed to tsunami-related coastal problems, which were mentioned in Strusińska and Oumeraci (2009b).
- It is believed to be more stable by a different grouping of the velocity time derivatives introduced in the momentum equation in comparison to the Wei et al. (1995) scheme (Lynett and Liu, 2002).
- It provides a good basis for the continuation of this study on the hydraulic performance of a submerged structure under tsunami conditions. The model features that can be used in the future work to improve the obtained results are: (i) coupling the COULWAVE and COBRAS models, (ii) improved model of wave run up, (iii) enhancement of the velocity field by means of the multi-layer approach (up to kh~6), and (iv) option of parallel computations by means of pCOULWAVE to reduce computational effort.

2.5 Specification of objectives and methodology

The objectives and methodology formulated tentatively in the introductory chapter are specified below on the basis of the present knowledge related to both tsunami modelling and hydraulic performance of low-crested structures.

2.5.1 Objectives

Application of an artificial reef as a countermeasure against tsunami impact might significantly enhance the effectiveness of the existing coastal protective systems. The incident tsunami energy reduced by the portion of the energy expected to be considerably dissipated through premature breaking in shallower water over the reef, might be more easily further attenuated by the successive defence lines, according to the concept of the gradual damping of extreme tsunami, as the wave continues propagating towards a shore (see Figure 1.1). However, the application of such submerged structures is at present restricted to storm waves, which in contrast to a tsunami are characterized by smaller periods, smaller velocities, smaller amount of transported energy and smaller wave shoaling.

Therefore, the aim of this study is to examine the hydraulic performance of an impermeable submerged structure under tsunami conditions, with particular consideration of the determination of both barrier dimensions required for the effective tsunami damping and barrier location, which govern the technical and economic feasibility of the structure. They can be summarized as follows:

- Identification of the global and local processes associated with the overall hydraulic performance of an impermeable submerged structure for tsunami-like solitary waves, on the basis of the results obtained from own laboratory experiments and numerical simulations.
- Improvement of the present knowledge on nonlinear transformation of a tsunami-like solitary wave over a submerged impermeable structure of a finite width on the basis of own laboratory experiments. The detailed characteristics of wave breaking and fission provided will be carefully confronted with the previous investigations on solitary wave transformation at obstacles of infinite width and regular/irregular wave interaction with various types of low-crested structures.
- Determination of damping performance of an artificial reef (wave transmission, wave reflection and wave energy dissipation) under varying incident tsunami conditions by means of a selected/improved available numerical model for water wave propagation, leading to the specification of reef dimensions corresponding to the most efficient tsunami attenuation.

2.5.2 Methodology

The overall methodology adopted in this study is illustrated by the flow chart in Figure 2.7. Since the required vertical-to-horizontal scale of a real long-wave tsunami is difficult to generate under laboratory conditions, the study will primarily be performed by means of a selected/improved numerical model for water wave propagation, and supplemented by own experimental results.

Considering the following aspects: (i) the required computational effort, (ii) the properties of wave theory implemented in the model, (iii) the availability of the computational model as an open source, (iv) its previous applications to water wave propagation and transformation problem, the computational code COULWAVE developed by Lynett (2002) was selected for this study.

The following working stages can be distinguished as shown in Figure 2.7:

- *Extensive literature review* (Chapter 2). It provides the knowledge essential to perform the successive stages of the study: (i) identification of the basic differences between a tsunami and storm waves, for which low-crested structures have been extensively applied, (ii) understanding of the processes associated with the hydraulic performance of low-crested structures, (iii) applicability of water wave theories and available numerical models to achieve the main objectives of this study.
- *Performance of own laboratory experiments* on solitary wave transformation at an impermeable submerged structure of a finite width (Chapter 3). In view of the research gaps related to the nonlinear interaction of a tsunami-like solitary wave with a submerged barrier of a finite width, well-design experiments must be performed, which aim at determination of wave breaking criterion, breaker types and number of solitons emerged in the fission process of a single solitary wave transmitted over the submerged barrier.
- *Selection, modification and validation of numerical model* (Chapter 4). On the basis of the results of the literature review, the numerical model for water wave propagation that is most appropriate for this study will finally be selected. Its validation using own experimental data and other available results is required to assure the correctness of the overall numerical analysis. Possible modifications/improvements will be introduced in the source code.
- *Numerical investigations of hydraulic performance of artificial reef for tsunami conditions* (Chapter 5). The analysis is performed by means of the selected numerical model for water wave propagation. Transmission, reflection and energy coefficients, expressed in terms of wave energy, are used to estimate the effectiveness of the structure to damp a tsunami. Wave energy losses due to breaking are only considered, while frictional losses

and energy dissipation due to vortex shedding are neglected. Due to time limitation of this study, implementation of wave energy dissipation caused by vortex generation was not possible. Therefore, this phenomenon is not modelled in this study; however it should be taken into account in the future work. The performed optimization analysis of reef geometry under varying incident solitary wave conditions results in the determination of the reef dimensions required for the most efficient tsunami attenuation.

- *Summary, conclusions and outlook* (Chapter 6). The key results of the entire study are summarized. Moreover, brief indications will be given on the feasibility of the proposed submerged structure, including recommendations for further investigations towards a practical implementation.

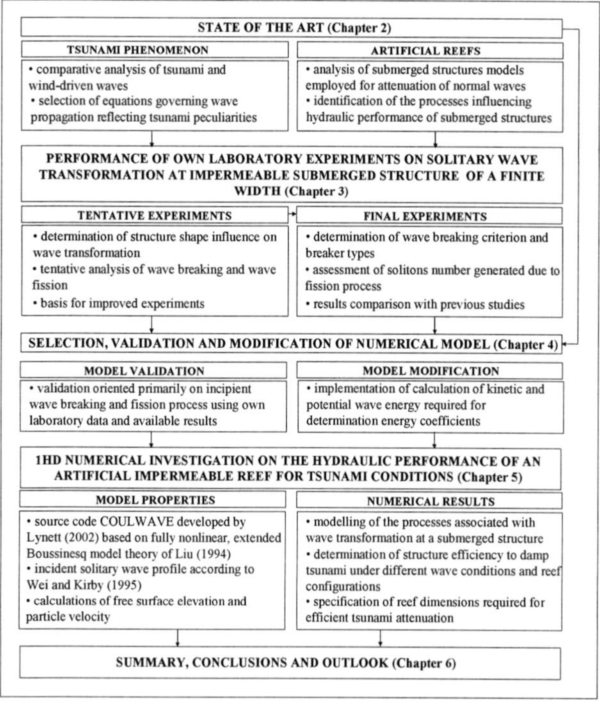

Figure 2.7: Detailed methodology adopted for this study

3. Laboratory experiments on nonlinear transformation of a solitary wave over a submerged structure of a finite width

As indicated by the extensive literature study discussed in Chapter 2, very little attention has been attracted to the investigation of solitary wave nonlinear transformation over low-crested structures of a finite width (including generation of wave breaking and wave fission) in comparison to their infinite-width counterparts. Since both these processes are expected to be induced over a submerged structure of a finite width, which is considered in this study, performance of own laboratory experiments is required in order to:

- establish a criterion for a solitary wave breaking over a submerged structure of a finite width,
- classify the breaker types developed in the process of solitary wave breaking,
- provide a detailed description of the fission mechanism, including determination of solitons number that emerged from a single transmitted solitary wave,
- compare the characteristics of the nonlinear processes with those reported in the previous studies for submerged barriers of infinite/finite widths under solitary/regular/irregular wave conditions,
- provide data for validation of the Boussinesq-type model COULWAVE.

More detailed information on the experimental programme and analysis of the laboratory results is provided in Strusińska and Oumeraci (2009e).

3.1 Experimental set-up

3.1.1 Wave flume

The laboratory experiments are performed in the wave flume of Leichtweiß-Institute for Hydraulic Engineering and Water Resources (LWI) at the Braunschweig Institute of Technology (Germany), which is approximately 90 m long, 2.0 m wide and 1.2 m high. A rubble-type wave absorber is placed at the end of the flume to minimize wave reflection from the end flume wall. The flume is of brick construction with two glass windows, each of approximately 10.0 m width, as illustrated in Figure 3.1.

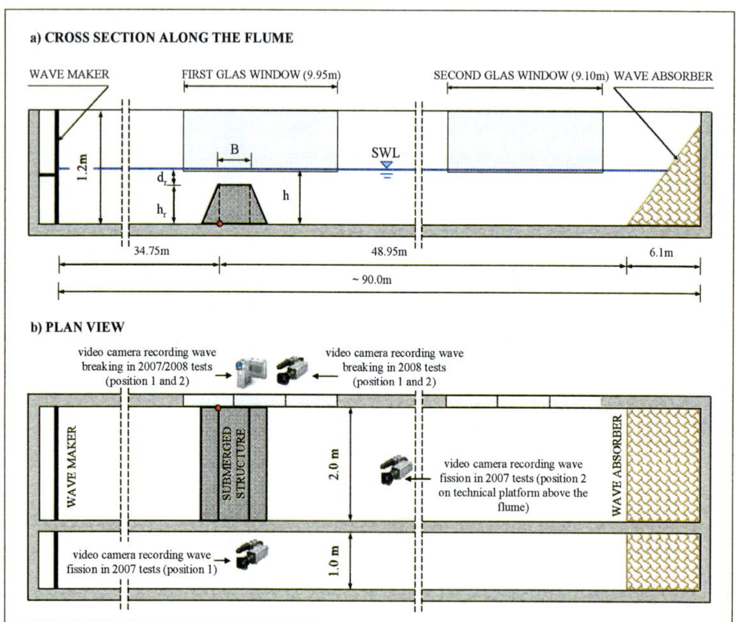

Figure 3.1: Model set-up in the wave flume of LWI in Braunschweig, Germany (figure not to scale)

3.1.2 Model of artificial reef

The geometry of the tested submerged structure, shown in Figure 3.2, is determined by its height h_r (or submergence depth d_r), crest width B as well as the gradient of the sea- and landward slopes.

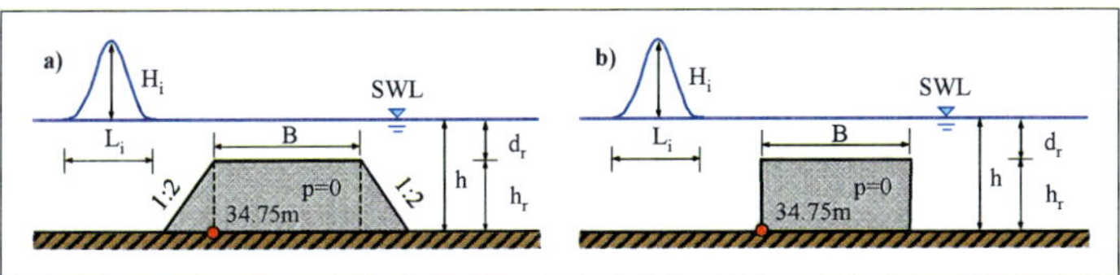

Figure 3.2: Definition sketch of structure geometry and incident wave conditions: a) trapezoidal reef, b) rectangular reef

The model of the artificial reef is represented in the experiments by a submerged impermeable barrier of a finite width, constructed as a steel frame consisting of two units each 1.0 m wide (see Figure 3.3a), covered by plywood plates (see Figure 3.3b), and additionally equipped with removable sea- and landward slopes 1:2 (see Figures 3.3c and d). This allows to investigate two barrier widths: B=1.0 m (when using a single unit only) and B=2.0 m (when using both units), each of trapezoidal (see Figures 3.2a, 3.3c) and rectangular shape (see Figures 3.2b and 3.3d). One of the units is fixed at the flume bottom at a distance of 34.75 m from the wave maker, which always corresponds to the top seaward structure corner, irrespective the barrier shape. The second unit, used to form the 2.0 m-wide structure, is placed behind the first unit and is screwed to the flume bottom, so that it needs only to be removed to perform tests with the 1.0 m wide reef.

By constructing the structure supports with removable elements, it is possible to adjust the barrier heights to the following values: h_r=0.3, 0.4, 0.5 m. The total water depth in front and behind the structure is varied (h=0.5, 0.6, 0.7 m) in order to examine different water depth conditions over the structure crest (d_r). The submergence depth d_r can be determined from the following relation (see Figure 3.2):

$$d_r = h - h_r \tag{3.1}$$

In order to examine the effect of the structure on the properties of progressive waves, additional experiments without the submerged barrier (i.e. for B=0.0 m and h_r=0.0 m) are conducted.

Figure 3.3: Construction details of submerged structures employed in experiments: a) steel-frame units, b) covering structure frame by plywood plates, c) trapezoidal barrier, d) rectangular barrier

3.1.3 Incident wave conditions

A tsunami is generated at different water depths h=0.5, 0.6, 0.7 m as a solitary wave of varying nominal incident wave height $H_{i,nom}$=0.06-0.22 m (with an interval of 0.02 m). The corresponding nominal incident wavelength $L_{i,nom}$ is approximated by means of the formula of Dean and Dalrymple (1991):

$$L_{i,nom} = 2\frac{2.12h}{\sqrt{H_{i,nom}/h}} \tag{3.2}$$

3.1.4 Experimental programme

The reef configurations and solitary wave incident conditions, employed in the laboratory experiments, are summarized in Table 3.1.

Table 3.1: Reef geometries and incident wave conditions tested

PARAMETER		UNIT	VALUE
Total water depth h=0.5m			
B	width of structure crest	[m]	1.0, 2.0
h_r	structure height	[m]	0.3, 0.4
d_r	submergence depth	[m]	0.2, 0.1
-	structure slopes	[-]	1:2
$H_{i,nom}$	nominal incident wave height	[m]	0.06, 0.08, 0.10, 0.12, 0.14, 0.16, 0.18, 0.20, 0.22
$L_{i,nom}$	nominal incident wavelength	[m]	6.12, 5.30, 4.74, 4.33, 4.01, 3.75, 3.53, 3.35, 3.2
Total water depth h=0.6m			
B	width of structure crest	[m]	0.0, 1.0, 2.0
h_r	structure height	[m]	0.0, 0.3, 0.4, 0.5
d_r	submergence depth	[m]	0.6, 0.3, 0.2, 0.1
-	structure slopes	[-]	1:2, no slopes
$H_{i,nom}$	nominal incident wave height	[m]	0.06, 0.08, 0.10, 0.12, 0.14, 0.16, 0.18, 0.20, 0.22
$L_{i,nom}$	nominal incident wavelength	[m]	8.04, 6.97, 6.23, 5.69, 5.27, 4.93, 4.64, 4.41, 4.20
Total water depth h=0.7m			
B	width of structure crest	[m]	1.0, 2.0
h_r	structure height	[m]	0.3, 0.4, 0.5
d_r	submergence depth	[m]	0.4, 0.3, 0.2
-	structure slopes	[-]	1:2
$H_{i,nom}$	nominal incident wave height	[m]	0.10, 0.12, 0.14, 0.16, 0.18, 0.20, 0.22
$L_{i,nom}$	nominal incident wavelength	[m]	7.85, 7.17, 6.64, 6.21, 5.85, 5.55, 5.29

The experimental programme is divided into two following parts:

- Experiments performed in 2007, termed hereafter the 2007 (preliminary) experiments: h=0.6 m, $H_{i,nom}$=0.06, 0.12, 0.18, 0.22 m, B=0.0, 1.0, 2.0 m, h_r=0.0, 0.3, 0.4, 0.5 m, trapezoidal (slopes 1:2) and rectangular structure.
- Experiments performed in 2008, termed hereafter the 2008 (detailed) experiments: h=0.5, 0.6, 0.7 m, $H_{i,nom}$=0.06-0.22 m (with an increment of

0.02 m), B=1.0, 2.0 m, h_r=0.3, 0.4, 0.5 m, trapezoidal structure (slopes 1:2).

The influence of the structure shape on the solitary wave transformation is determined from the results of the 2007 experiments. Moreover, these tests provide a basis for the performance of more detailed laboratory investigations (i.e. the 2008 experiments) on solitary wave breaking conditions and generation/evolution of the fission phenomenon. The entire experimental programme is summarized in Table A1 for the preliminary experiments and Table A2 for the detailed experiments in Appendix A.

3.1.5 Measurements and observation technique

Free surface elevation of a solitary wave is measured by means of wire-resistance wave gauges, whose position is adjusted in each new test to record accurately the inception of wave breaking and generation/evolution of the disintegration of a single wave into a solitary wave train. In comparison to the tentative 2007 tests, a greater number of wave gauges is required in the detailed 2008 experiments, as exemplarily illustrated in Figure 3.4.

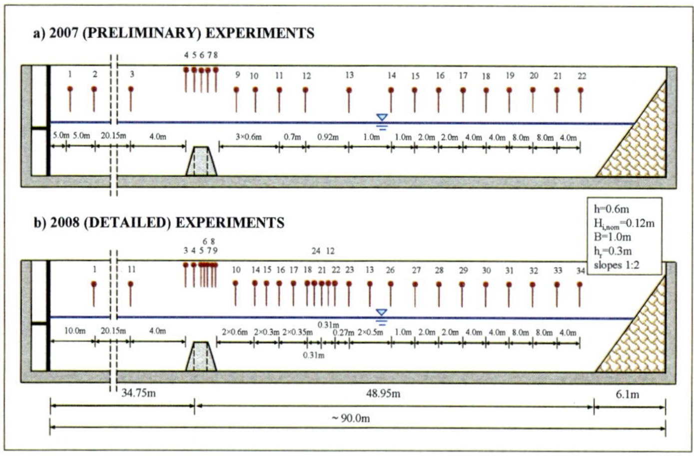

Figure 3.4: Exemplary wave gauge arrangement for following conditions: h=0.6 m, $H_{i,nom}$=0.12 m, B=1.0 m, h_r=0.3 m, trapezoidal reef with slopes 1:2: a) in preliminary tests 2007, b) in detailed tests 2008 (figure not to scale)

Additionally, two video cameras are used to record the incipient wave breaking, the development of the wave breakers as well as the propagation of the solitary wave train emerged due to wave scattering by the submerged structure. The positions of the cameras in the 2007 and 2008 tests are shown in Figure 3.1.

3.2 Analysis of the experimental results

3.2.1 Definition of dimensionless variables

For the purposes of the data analysis, the reef geometries and the solitary wave incident conditions are expressed in terms of dimensionless ratios, provided in Tables 3.2 and 3.3, respectively for the nominal incident wave height $H_{i,nom}$ (as given in Table 3.1).

Table 3.2: Dimensionless reef geometry tested in 2007 and 2008 experiments for nominal incident wave height $H_{i,nom}$ (see definition in Figure 3.2)

PARAMETER		UNIT	VALUE			
Total water depth h=0.5m			h_r=0.0m d_r=0.5m	h_r=0.3m d_r=0.2m	h_r=0.4m d_r=0.1m	h_r=0.5m d_r=0.0m
h_r/h	relative structure height	[-]	-	0.6	0.8	-
d_r/h	relative submergence depth	[-]	-	0.4	0.2	-
Total water depth h=0.6m			h_r=0.0m d_r=0.6m	h_r=0.3m d_r=0.3m	h_r=0.4m d_r=0.2m	h_r=0.5m d_r=0.1m
h_r/h	relative structure height	[-]	0.0	0.5	0.67	0.83
d_r/h	relative submergence depth	[-]	1.0	0.5	0.33	0.17
Total water depth h=0.7m			h_r=0.0m d_r=0.7m	h_r=0.3m d_r=0.4m	h_r=0.4m d_r=0.3m	h_r=0.5m d_r=0.2m
h_r/h	relative structure height	[-]	-	0.43	0.57	0.71
d_r/h	relative submergence depth	[-]	-	0.57	0.43	0.29

3.2.2 Determination of incident wave conditions

Since the generated solitary wave profiles differs from the theoretical solitary wave solution, for which the entire wave profile lies above still water level (i.e. there is no wave trough), it is necessary to redefine the incident wave height. For this purpose three definitions might be considered (see Figure 3.5):

Table 3.3: Dimensionless incident solitary wave conditions tested in 2007 and 2008 experiments ($H_i=H_{i,nom}$, $L_i=L_{i,nom}$)

PARAMETER			H_i=0.06m L_i=6.12m	H_i=0.08m L_i=5.30m	H_i=0.10m L_i=4.74m	H_i=0.12m L_i=4.33m	H_i=0.14m L_i=4.01m	H_i=0.16m L_i=3.75m	H_i=0.18m L_i=3.53m	H_i=0.20m L_i=3.35m	H_i=0.22m L_i=3.20m
rel. inc. wave height	**H_i/h [-]**	**h=0.5m**	0.12	0.16	0.20	0.24	0.28	0.32	0.36	0.40	0.44
	H_i/d_r [-]	d_r=0.2m	0.30	0.40	0.50	0.60	0.70	0.80	0.90	1.00	1.10
		d_r=0.1m	0.60	0.80	1.00	1.20	1.40	1.60	1.80	2.00	2.20
rel. crest width	B/L_i [-]	B=1.0m	0.16	0.19	0.21	0.23	0.25	0.27	0.28	0.30	0.31
		B=2.0m	0.33	0.38	0.42	0.46	0.50	0.53	0.57	0.60	0.63
PARAMETER			H_i=0.06m L_i=8.04m	H_i=0.08m L_i=6.97m	H_i=0.10m L_i=6.23m	H_i=0.12m L_i=5.69m	H_i=0.14m L_i=5.27m	H_i=0.16m L_i=4.93m	H_i=0.18m L_i=4.64m	H_i=0.20m L_i=4.41m	H_i=0.22m L_i=4.20m
rel. inc. wave height	**H_i/h [-]**	**h=0.6m**	0.10	0.13	0.17	0.20	0.23	0.27	0.30	0.33	0.37
	H_i/d_r [-]	d_r=0.3m	0.20	0.27	0.33	0.40	0.47	0.53	0.60	0.67	0.73
		d_r=0.2m	0.30	0.40	0.50	0.60	0.70	0.80	0.90	1.00	1.10
		d_r=0.1m	0.60	0.80	1.00	1.20	1.40	1.60	1.80	2.00	2.20
rel. crest width	B/L_i [-]	B=1.0m	0.12	0.14	0.16	0.18	0.19	0.20	0.22	0.23	0.24
		B=2.0m	0.25	0.29	0.32	0.35	0.38	0.41	0.43	0.45	0.48
PARAMETER			H_i=0.06m L_i=10.14m	H_i=0.08 m L_i=8.78 m	H_i=0.10m L_i=7.85m	H_i=0.12m L_i=7.17m	H_i=0.14m L_i=6.64m	H_i=0.16m L_i=6.21m	H_i=0.18m L_i=5.85m	H_i=0.20m L_i=5.55m	H_i=0.22m L_i=5.29m
rel. inc. wave height	**H_i/h [-]**	**h=0.7m**	-	-	0.14	0.17	0.20	0.23	0.26	0.29	0.31
	H_i/d_r [-]	d_r=0.4m	-	-	0.25	0.30	0.35	0.40	0.45	0.50	0.55
		d_r=0.3m	-	-	0.33	0.40	0.47	0.53	0.60	0.67	0.73
		d_r=0.2m	-	-	0.50	0.60	0.70	0.80	0.90	1.00	1.10
rel. crest width	B/L_i [-]	B=1.0m	-	-	0.13	0.14	0.15	0.16	0.17	0.18	0.19
		B=2.0m	-	-	0.25	0.28	0.30	0.32	0.34	0.36	0.38

Incident wave height $H_{i,1}$ measured from the actual still water level to the wave crest, corresponding to the wave height determined according to the solitary wave theory.

- Incident wave height $H_{i,2}$ as the height measured from the actual still water level to the wave crest plus half of the height between the still water level and the wave trough appearing at the rear wave part:

$$H_{i,2} = H_{i,1} + 0.5\left(H_{i,3} - H_{i,1}\right) \tag{3.3}$$

- Incident wave height $H_{i,3}$ measured from the wave trough, appearing at the rear wave part, to the wave crest.

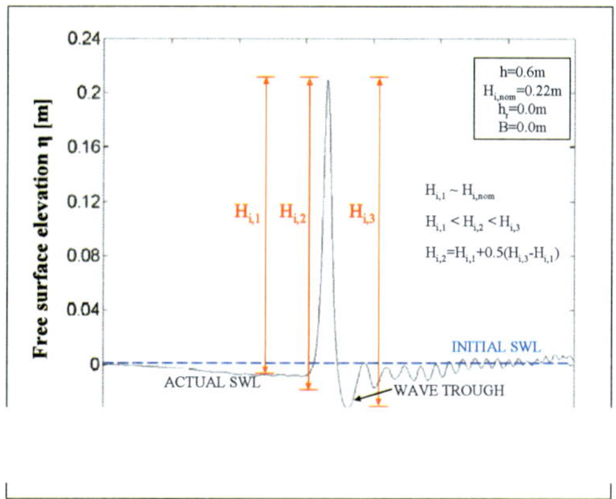

Figure 3.5: Definition of incident solitary wave height in experiments

The incident solitary wave heights $H_{i,1}$, $H_{i,2}$ and $H_{i,3}$ are determined as the averaged values on the basis of the 2007 and 2008 experiments performed for the trapezoidal structure (for details see Strusińska and Oumeraci, 2009e). Their values are compared with the nominal incident wave height $H_{i,nom}$ in Table 3.4.

The discrepancies between the nominal and the measured incident wave parameters (the incident wave height H_i and incident wavelength L_i) are plotted in Figure 3.6. Generally, incident wave properties for the wave of height $H_{i,1}$ correspond to the nominal incident wave parameters.

The incident wave characteristics for the incident wave heights $H_{i,1}$, $H_{i,2}$ and

$H_{i,3}$, given in terms of dimensional and dimensionless parameters, are provided in Tables B1-B3, respectively in Appendix B.

Table 3.4: Determined incident solitary wave heights in experiments

$H_{i,nom}$ [m]	$H_{i,1}$ [m]	$H_{i,2}$ [m]	$H_{i,3}$ [m]
0.060	0.060	0.060	0.061
0.080	0.082	0.085	0.089
0.100	0.103	0.106	0.110
0.120	0.122	0.127	0.133
0.140	0.143	0.152	0.161
0.160	0.166	0.176	0.187
0.180	0.180	0.191	0.201
0.200	0.205	0.219	0.234
0.220	0.224	0.237	0.299

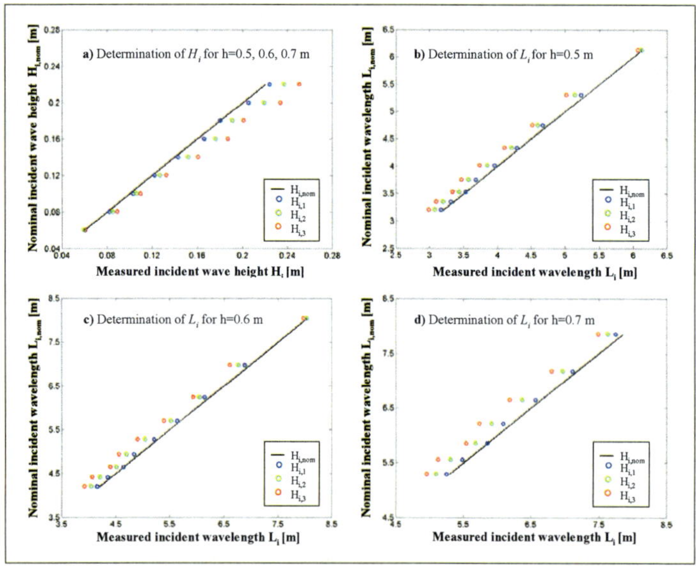

Figure 3.6: Comparison of nominal and measured incident wave parameters: a) wave heights for water depth h=0.5-0.7 m, b) wavelengths for h=0.5 m, c) wavelengths for h=0.6 m, d) wavelengths for h=0.7 m

3.2.3 Effect of structure shape on breaking conditions

The effect of the shape of the submerged structure on the nonlinear transformation of incident solitary waves is examined in the 2007 experiments, in which trapezoidal (slopes 1:2) and rectangular submerged barriers are used. The results, plotted in Figure 3.7 and summarized in Tables C1 and C2 in Appendix C, indicate that the barrier shape does not affect the inception of wave breaking and the wave breaker types for the tested conditions (Strusińska and Oumeraci, 2008).

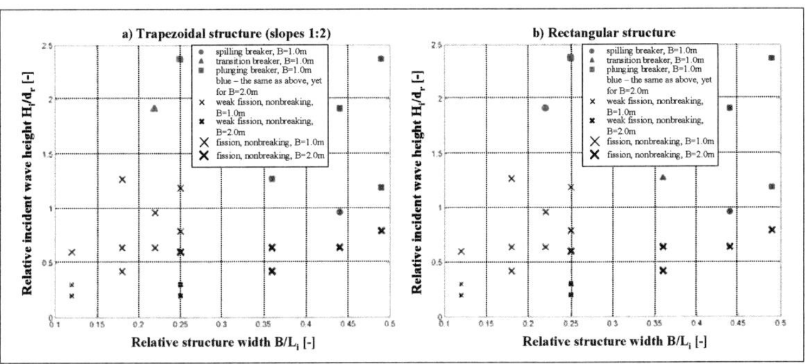

Figure 3.7: Effect of structure slope on wave breaking characteristic: a) trapezoidal structure, b) rectangular structure (H_i=$H_{i,2}$)

Since the effect of the structure shape on the solitary wave transformation is found to be negligible for the tested conditions, the 2008 (detailed) experiments are performed for the trapezoidal barrier only.

3.2.4 Classification of solitary wave evolution modes

The following evolution modes of a solitary wave transmitted over a submerged structure of a finite width are determined on the basis of the 2007 and 2008 laboratory experiments (see Tables C1-C4 in Appendix C):

- *No fission of a single incident nonbreaking wave* observed for tests without the barrier, i.e. for B=0.0 m and h_r=0.0 m (see Figure 3.8).
- *Weak distortion of a single transmitted nonbreaking wave* characterized by a partial development of the second soliton in case of weakly nonlinear waves

($H_{i,nom}/h$=0.1) and relatively large submergence depths d_r/h=0.5 and 0.33 for both structure widths B=1.0 and B=2.0 m. This is due to the relatively late inception of the fission phenomenon in comparison to the other considered cases, which cannot be fully observed due to the insufficiently long wave flume. The emerged solitons are trailed by oscillatory waves. Additionally, no wave breaking of the leading soliton is generated (see Figure 3.9).

- *Fission of a single transmitted nonbreaking wave* with the incident wave disintegrating into two and more solitons, trailed by oscillatory waves. No breaking of the leading soliton is observed (see Figure 3.10).
- *Fission of the breaking leading wave* which can be further divided into the following cases, depending on the order of the inception of the nonlinear phenomena:
  - *Fission of a single transmitted wave followed by the breaking of the leading soliton.* This mode of wave evolution is observed for wave breaking induced behind the submerged structure – the single transmitted wave tends to disintegrate into solitons first, what is indicated by small humps appearing on the rear part of the wave profile. However, further development of wave fission is interrupted by the inception of breaking of the main wave. Once the breaking event is completed, fission of the broken wave continues till the full development of the successive solitons, followed by oscillatory waves (see Figure 3.11).
  - *Breaking of a single transmitted wave followed by wave fission of the broken wave.* This wave behaviour is typical for the location of incipient breaking shifted towards the landward corner of the structure, what occurs for higher relative incident waves H_i/h, decreasing relative submergence depth d_r/h and increasing relative structure width B/L_i. The single transmitted wave breaks first and then disintegrates into a train of solitons, trailed by oscillatory waves (see Figure 3.12).

However, the two aforementioned evolution modes are difficult to distinguish due to the complexity of solitary wave transformation over the structure. In the 2007 tests, only the wave disintegration accompanied by wave breaking is investigated. In the 2008 experiments, however, both phenomena are analysed separately, with the fission process developed under nonbreaking wave conditions only. Therefore, more detailed experiments would be required to confirm the phenomenon of wave fission followed by breaking.

Incident wave conditions leading to the generation of wave breaking in front of the structure, as examined by Losada et al. (1989) and Liu and Cheng (2001), are not considered in this study. The results of own experiments, together with the study of Liu and Cheng (2001), are in contrast with the observations of Losada et al. (1989), who postulated that wave breaking does not suppress the fission process, only if the breaking event occurs before the fission mechanism started.

The leading waves break generally as spilling and plunging breakers; detailed characteristics of the breakers are provided in the next section. Unlike Losada et al. (1989) and Liu and Cheng (2001), the broken wave does not propagate as an undular bore, yet as a stable solitary wave.

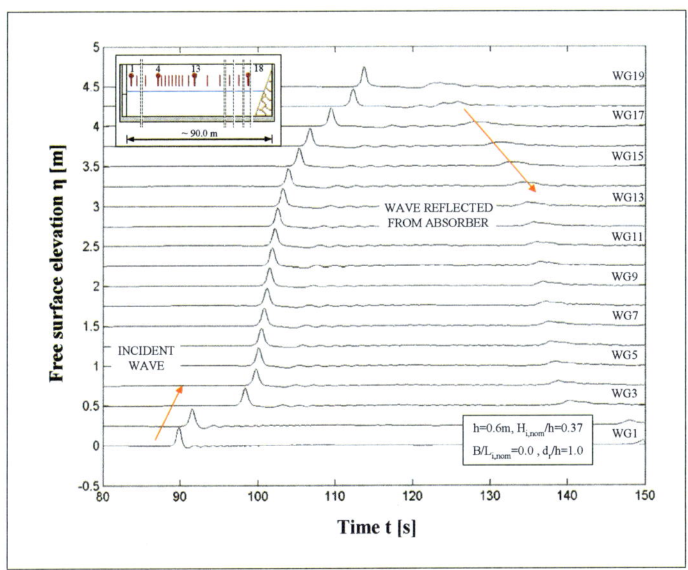

Figure 3.8: No fission of single incident nonbreaking solitary wave observed exemplarily for $H_{i,nom}/h=0.37$, $B/L_{i,nom}=0.0$, $d_r/h=1.0$, h=0.6 m

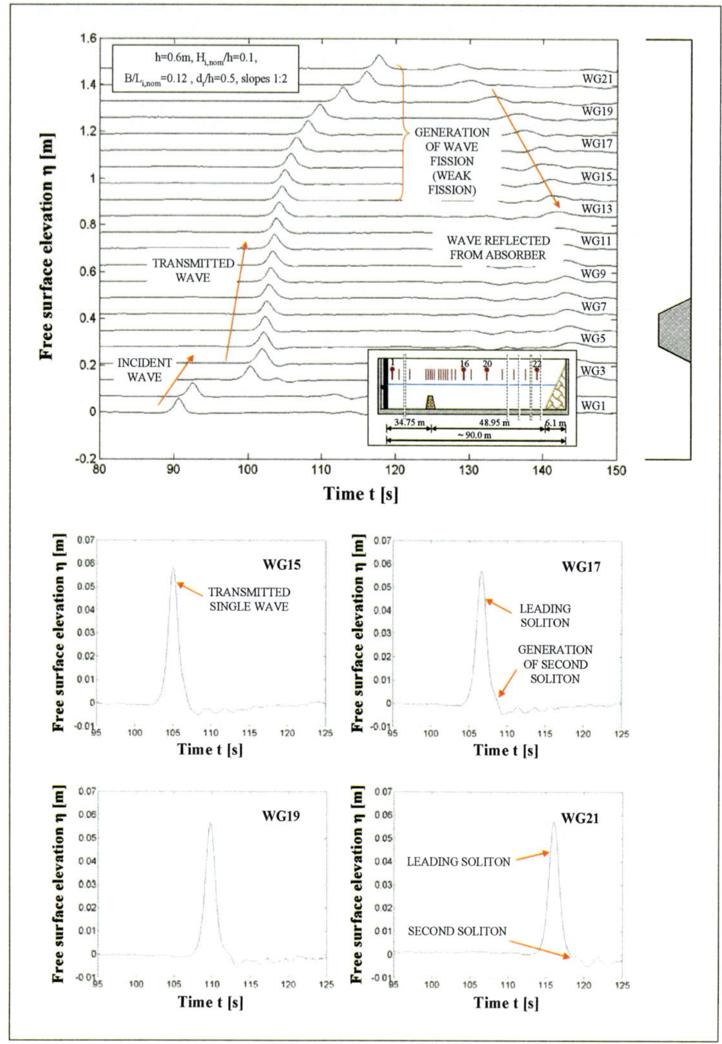

Figure 3.9: Weak distortion of single transmitted nonbreaking solitary wave observed exemplarily for $H_{i,nom}/h=0.1$, $B/L_{i,nom}=0.12$, $d_r/h=0.5$, h=0.6 m, B=1.0 m

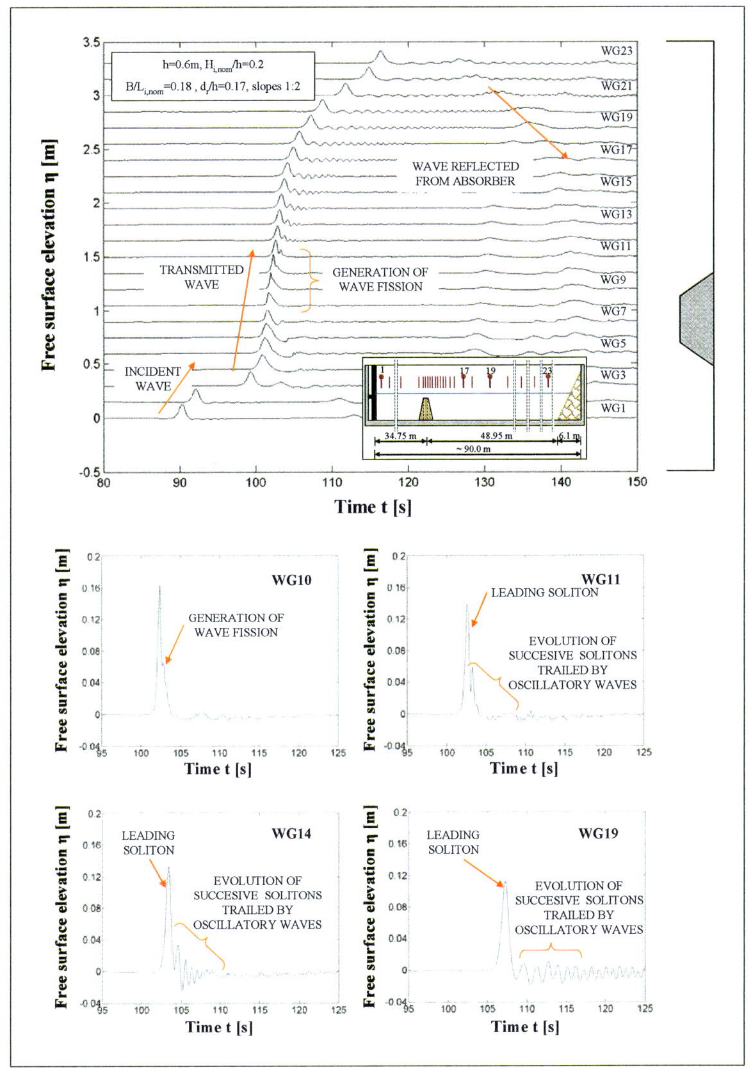

Figure 3.10: Fission of single transmitted nonbreaking solitary wave observed exemplarily for $H_{i,nom}/h=0.2$, $B/L_{i,nom}=0.18$, $d_r/h=0.17$, h=0.6 m, B=1.0 m, reef slopes 1:2

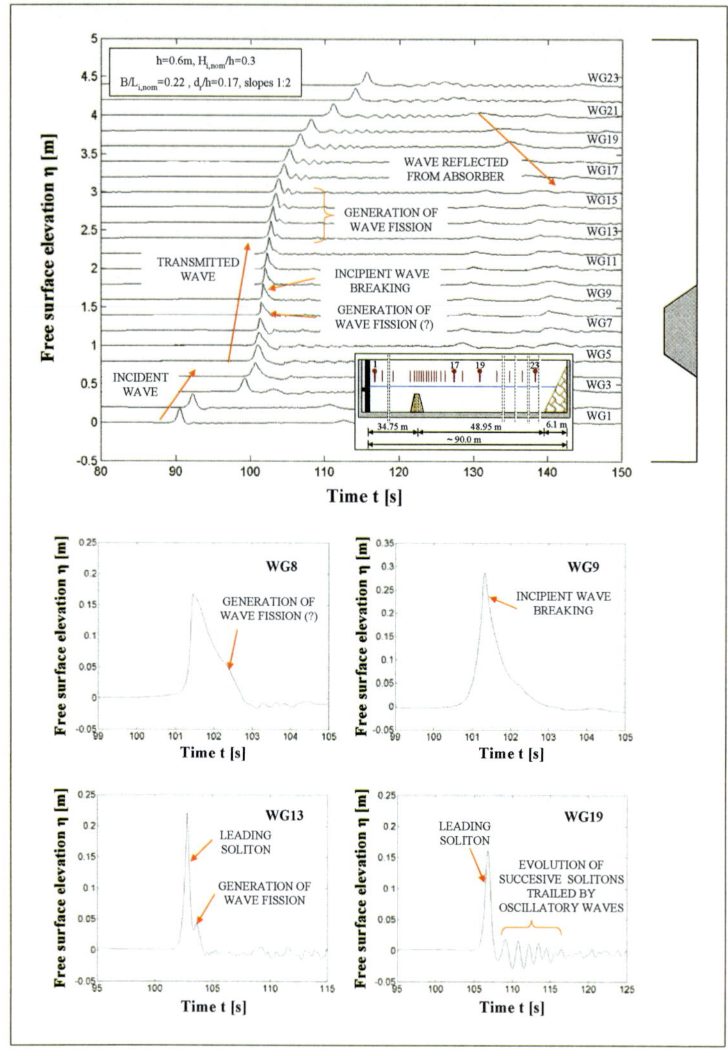

Figure 3.11: Fission of single transmitted solitary wave followed by breaking observed exemplarily for $H_{i,nom}/h=0.30$, $B/L_{i,nom}=0.22$, $d_r/h=0.17$, h=0.6 m, B=1.0 m, reef slopes 1:2

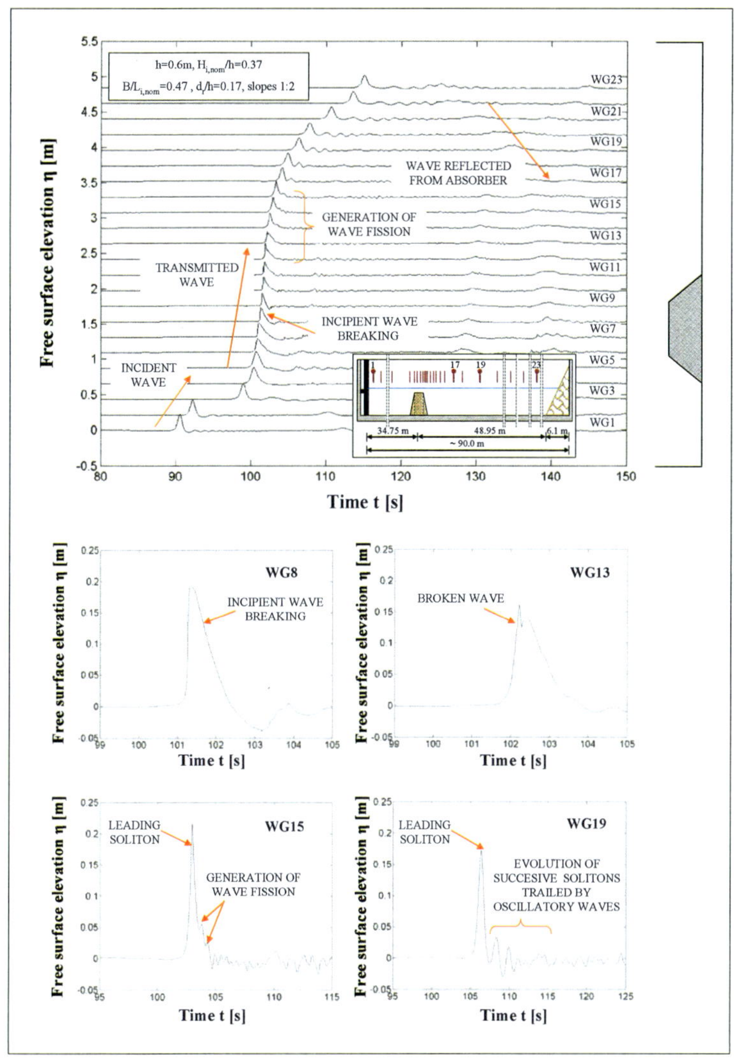

Figure 3.12: Breaking of single transmitted solitary wave followed by fission observed exemplarily for $H_i/h=0.37$, $B/L_{i,nom}=0.47$, $d_r/h=0.17$, h=0.6 m, B=2.0 m, reef slopes 1:2

3.2.5 Classification of breaker types

On the basis of the 2007 and 2008 experiments, the breaker types developed in case of a solitary wave breaking over submerged structures of a finite width are classified as follows:

- *Spilling breaker* with white water appearing at wave crest and expanding gradually with the propagation distance towards the frontal wave foot. It is generally observed for weakly nonlinear waves and large relative submergence depths d_r/h (see Figure 3.13).
- *Transition breaker* representing a transition form between spilling and plunging breakers.
- *Plunging breaker* with the wave crest overturning and plunging as a jet on the frontal wave foot, generating eventually a splash. It is generally observed for nonlinear waves and small relative submergence depths d_r/h (see Figure 3.14).

Unlike the breaker types reported by Bleck (2003) for periodic waves breaking over a rectangular submerged structure with a finite width, the breaker formations observed in the experiments correspond to those developed over natural sloping beaches (Patrick and Wiegel, 1954). They also confirm the classification introduced by Cooker et al. (1990) and Grilli et al. (1994) for the forward breaking of a solitary wave over finite-width submerged barriers (as discussed in Section 2.3.3). However, the additional breaker type, termed by Cooker et al. (1990) and Grilli et al. (1994) the backward breaking of the wave tail in the direction opposite to the propagation of the main wave, is not observed in the experiments performed in this study.

3.2.6 Determination of breaking criterion

On the basis of the results of the preliminary 2007 experiments (performed under the following conditions: h=0.6 m; $H_{i,nom}$=0.06, 0.12, 0.18, 0.22 m; B=0.0, 1.0, 2.0 m; h_r=0.0, 0.3, 0.4, 0.5 m; trapezoidal and rectangular reef), a negligible effect of the structure shape on the solitary wave breaking characteristic is found (see Section 3.2.3).

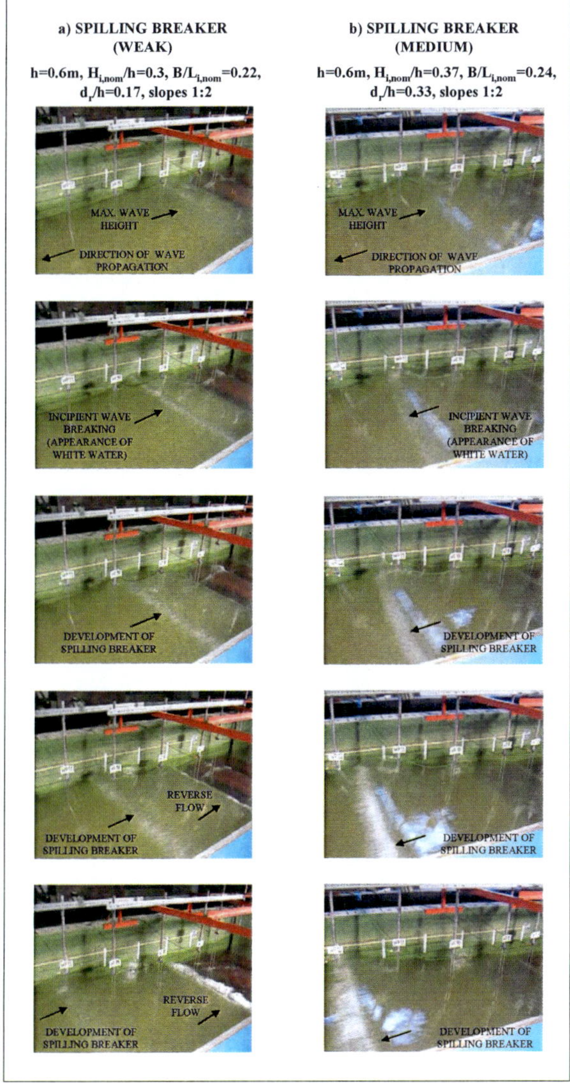

Figure 3.13: Development of spilling wave breaker: a) with weak rate of wave energy dissipation, b) with medium rate of wave energy dissipation

Figure 3.14: Development of plunging wave breaker: a) with weak rate of wave energy dissipation, b) with medium rate of wave energy dissipation, c) with strong rate of wave energy dissipation

Therefore, the criterion for a solitary wave breaking over a submerged barrier of a finite width is determined for a trapezoidal-shaped structure, employed in the detailed 2008 experiments (h=0.5, 0.6, 0.7 m; $H_{i,nom}$=0.06-0.22 m; B=1.0, 2.0 m; h_r=0.3, 0.4, 0.5 m). The breaking limit can be however applied to both trapezoidal and rectangular barriers.

The detailed breaking characteristics in the 2007 and 2008 laboratory experiments is provided in Tables C3 and C4 in Appendix C for the trapezoidal structure of width B=1.0 m and B=2.0 m, respectively.

The breaking criterion, expressed in terms of the incident solitary wave height $H_{i,2}$, is valid within the range B/L_i=0.2-0.49; H_i/d_r=0.2-2.37 and yields:

$$H_i/d_r = 0.573\ (B/L_i)^{-0.6} \tag{3.4}$$

By employing the incident wave height $H_{i,1}$ (corresponding generally to the nominal incident wave height $H_{i,nom}$), the lower boundary of the breaking limit is obtained (see Figure 3.15):

$$H_i/d_r = 0.544\ (B/L_i)^{-0.58} \tag{3.5}$$

Similarly, the upper boundary of the breaking criterion is determined for the incident solitary wave height $H_{i,3}$ (see Figure 3.15):

$$H_i/d_r = 0.626\ (B/L_i)^{-0.58} \tag{3.6}$$

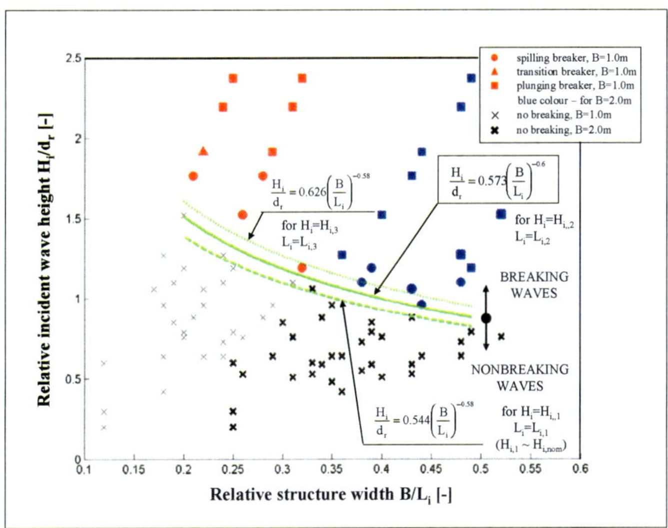

Figure 3.15: Criterion for a solitary wave breaking over a submerged structure of a finite width

The aforementioned criteria were determined by fitting a line to a set of data containing points representing the transition from the nonbreaking to breaking waves (i.e. points from the upper domain of nonbreaking waves and points from the lower domain of breaking waves).

Figure 3.15 shows a clear tendency of the breaking limit to reach a constant value for the relative structure width $B/L_i>0.5$. A similar behaviour was reported in numerical studies of Iwata et al. (1996) on breaking of periodic waves over a rectangular structure (see Strusińska and Oumeraci, 2009b).

3.2.7 Description of solitary wave fission

The analysis of the process of solitary wave fission over a submerged structure, preliminary investigated in the 2007 tests under nonbreaking and breaking wave conditions, is supplemented by the results of the 2008 experiments, in which only nonbreaking waves are considered (for h=0.6 m, h_r=0.3, 0.4, 0.5 m, B=1.0 and 2.0 m, trapezoidal structure). The analysis includes the determination of the number of the solitons emerged from a single transmitted solitary wave as well as spatial evolution of the height of the leading wave.

In order to distinguish between the solitons and oscillatory waves trailing the solitons, very detailed measurements of free surface elevation are required, what is achieved by locating wave gauges very densely within the region of wave scattering. With respect to the lack of guidelines for the prediction of the soliton number in case of submerged structures of a finite width, it is assumed that wave fission is a chain process, and thus the solitons are generated successively (i.e. leading wave considered as first soliton, second, third, ..., j-th soliton).

The results of the analysis of the fission phenomenon on the basis of the 2007 and 2008 experiments are summarized in Tables D1 and D2 in Appendix D, for trapezoidal structure of widths B=1.0 m and B=2.0 m, respectively. It should be underlined, that the leading wave always represents the first soliton.

Similarly to Lin (2004), only the leading soliton tends to resemble a solitary wave, while the successively emerged solitons resemble rather oscillatory waves.

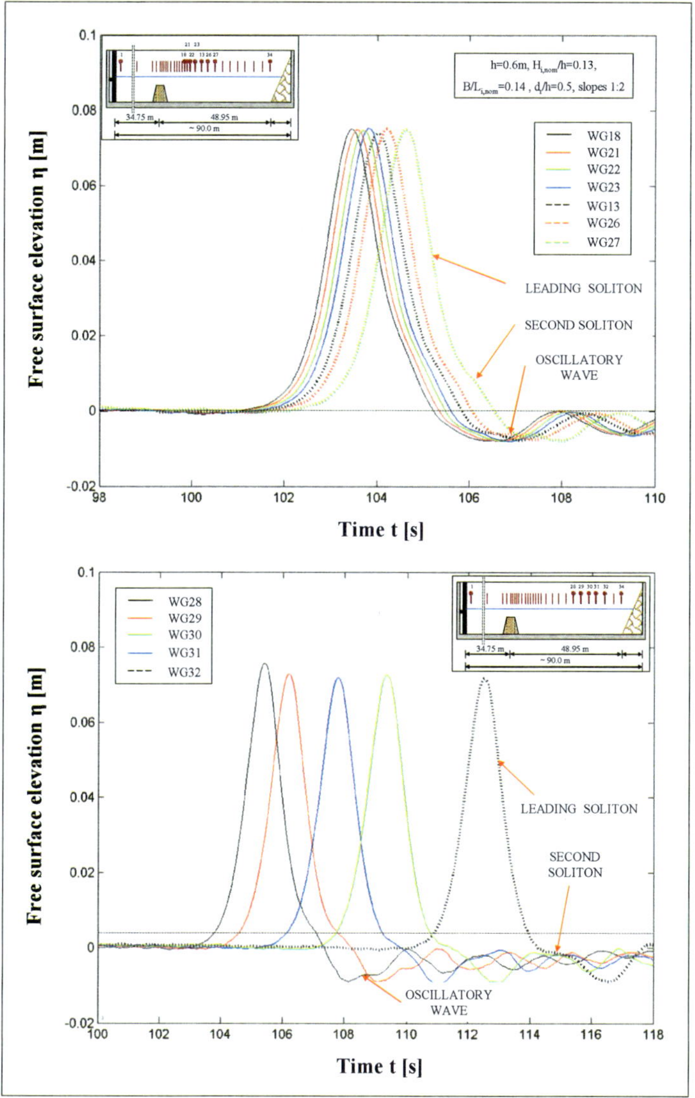

Figure 3.16: Generation and evolution of solitons for $H_{i,nom}/h$=0.13, $B/L_{i,nom}$=0.14, d_r/h=0.5, h=0.6 m, B=1.0 m, reef slopes 1:2 (number of solitons N=2)

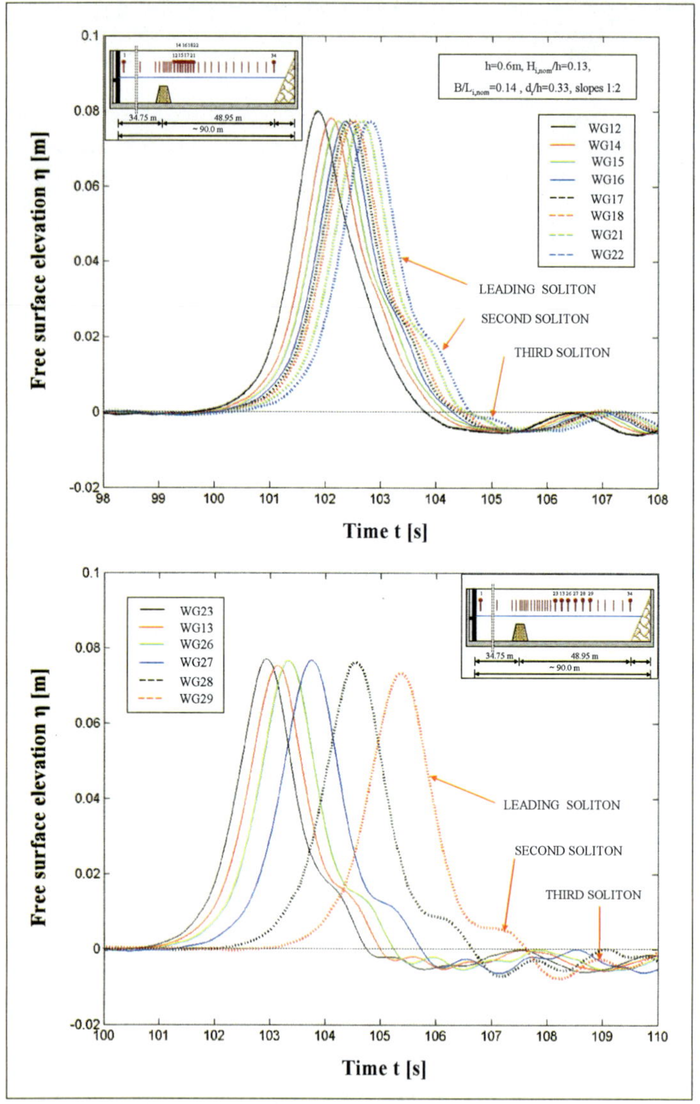

Figure 3.17: Generation and evolution of solitons for $H_{i,nom}/h$=0.13, $B/L_{i,nom}$=0.14, d_r/h=0.33, h=0.6 m, B=1.0 m (number of solitons N=3)

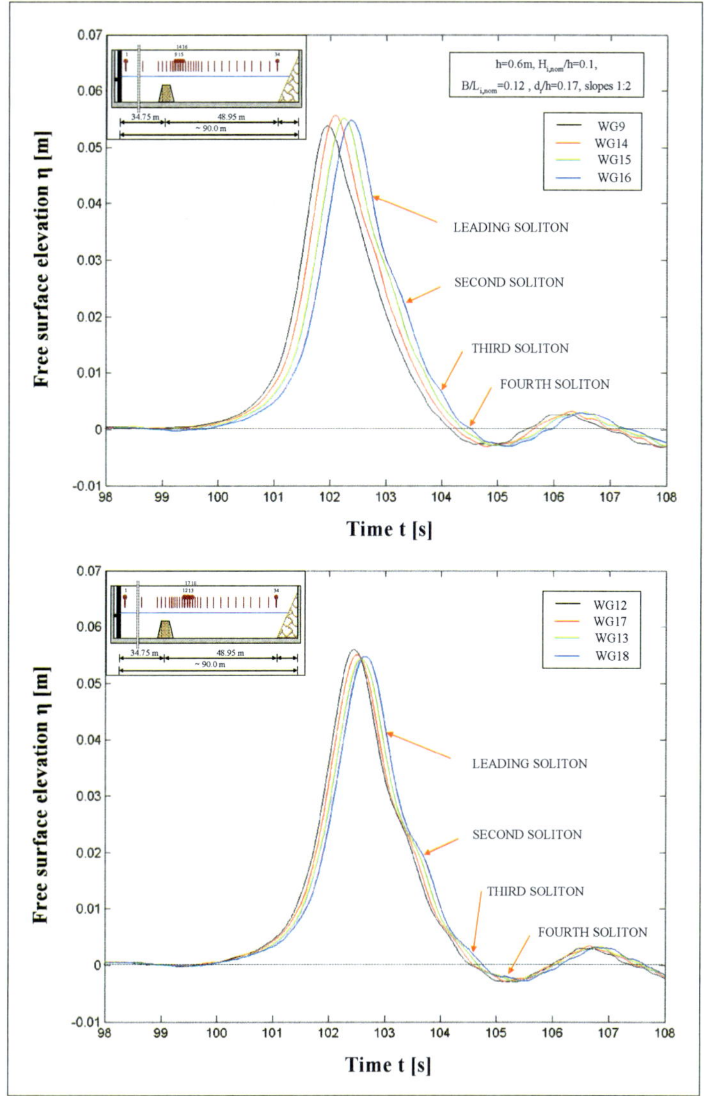

Figure 3.18: Generation and evolution of solitons for $H_{i,nom}/h=0.1$, $B/L_{i,nom}=0.12$, $d_r/h=0.17$, h=0.6 m, B=1.0 m (number of solitons N=4)

The number of the emerged solitons, as reported for the case of solitary wave scattering by submerged structures of infinite widths, is merely dependent on relative water depth over the barrier d_r/h. However, the experiments indicate the additional influence of relative structure width B/L_i on the number of the waves in the solitary wave train. The soliton number tends to be constant for a given ratio d_r/h, irrespective incident wave height H_i/h, and increases as the water over the obstacle crest becomes shallower, as shown in Figure 3.19.

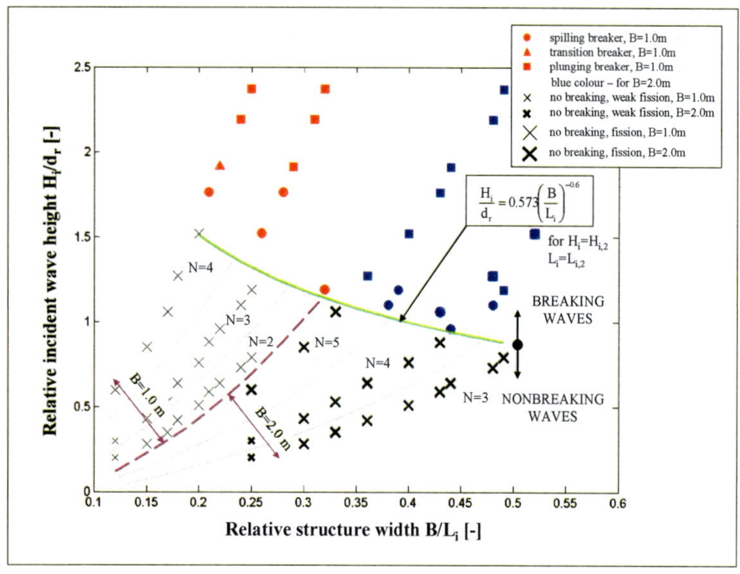

Figure 3.19: Modes of solitary wave evolution over submerged trapezoidal structure of widths B=1.0 m and 2.0 m ($H_i=H_{i,2}$, $L_i=L_{i,2}$)

In case of shorter structures (B=1.0 m and $B/L_{i,nom}$=0.12-0.24), the observed soliton number is: N=2 for d_r/h=0.5 (see Figure 3.16), N=3 for d_r/h=0.33 (see Figure 3.17) and N=4 for d_r/h=0.17 (see Figure 3.18). Widening of the barrier width to B=2.0 m ($B/L_{i,nom}$=0.25-0.48) results in the increase of the soliton number by one in comparison to the shorter barriers: N=3 for d_r/h=0.5, N=4 for d_r/h=0.33 and N=5 for d_r/h=0.17. It should however be noted that the method used for the determination of the soliton number requires further improvements in order to accurately distinguish between the emerged solitons and the oscillatory wave train.

Additionally, spatial evolution of maximum free surface elevation (corresponding to the wave height of the single incident wave/leading soliton) is analysed. It should be noted that the first wave gauge is placed in the experiments at the distance of 10.0 m from the wave maker.

In case of nonbreaking waves, the reduction of the incident wave is caused by wave reflection from the seaward face of the structure, as shown exemplarily in Figure 3.20. The rate of wave transmission decreases with decreasing relative water depth over the barrier crest d_r/h.

Significant incident wave attenuation can be observed for waves breaking at the structure, as presented exemplarily in Figure 3.21. The minimum wave transmission (about 72%) corresponds to larger incident wave heights passing the wider structure of the smallest submergence depth. The heights of the broken leading wave tend to be constant once the breaking event is completed.

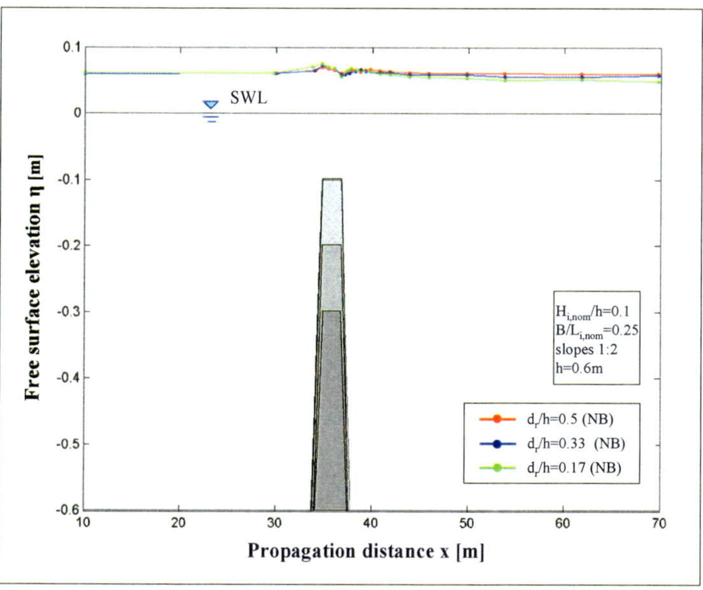

Figure 3.20: Spatial evolution of maximum surface elevation for $H_{i,nom}/h$=0.1, $B/L_{i,nom}$=0.25, h=0.6 m, B=2.0 m, reef slopes 1:2

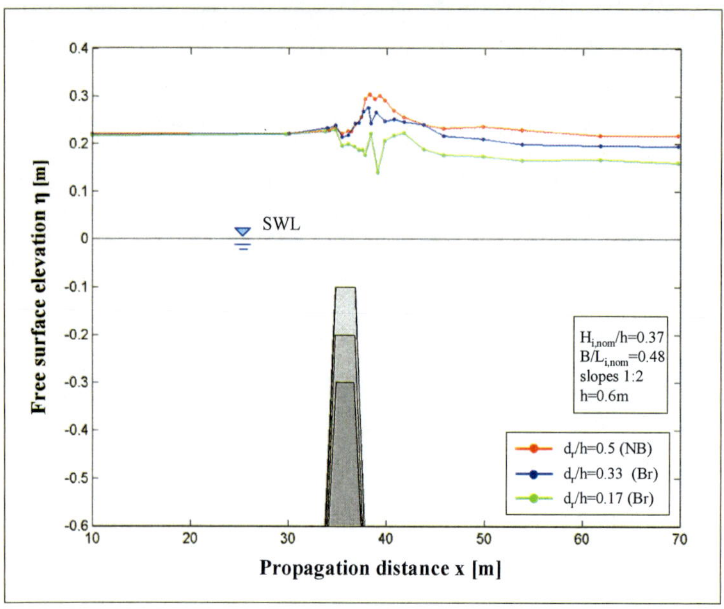

Figure 3.21: Spatial evolution of maximum surface elevation for $H_{i,nom}/h$=0.37, $B/L_{i,nom}$=0.48, h=0.6 m, B=2.0 m, reef slopes 1:2

The solitons number emerged due to wave scattering by the structure of a finite width is compared with that reported for solitary wave fission over submerged structures of infinite widths (see Table 3.5). In the latter case, there are several available theoretical formulations allowing to predict the number of the generated solitons, given the water depth conditions behind and over the barrier, e.g. Johnson (1972) given by Eq. (2.63) or Germain (1984) and Kabbaj (1985) given by Eq. (2.64).

The investigations used in the comparative analysis of the soliton number are those by Madsen and Mei (1969), Johnson (1972) and Seabra-Santos et al. (1987). The theory of Johnson (1972) was verified with the numerical results of Madsen and Mei (1969), while Seabra-Santos et al. (1987) applied the formula of Germain (1984) and Kabbaj (1985) to confirm the soliton number generated under laboratory conditions for both nonbreaking and breaking waves.

The analysis of the results indicates that the structure of a larger finite width (B=2.0 m) tends to act in a similar manner to the barriers of infinite widths for larger relative submergence depths d_r/h. In case of smaller ratios d_r/h, there is however no available data for the infinite structures to be compared with those related to the structures with a finite width.

Moreover, discrepancies are found between the solitons number reported in the available studies and that predicted theoretically (by simply inserting the considered ratio d_r/h into the equations).

Finally, differences between the available theories can be observed; the formula provided by Johnson (1972) tends generally to overestimate the solitons number in comparison to the approach of Germain (1984) and Kabbaj (1985). Since both theories are verified by means of available studies, it is difficult to recognize which of them predicts the soliton numbers correctly.

Table 3.5: Comparison of solitons number generated over submerged structures of finite and infinite widths

d_r/h [-]	Obstacles of finite widths (own experiments)		Obstacles of infinite widths (B→∞)				
			Theoretical formulations *		Previous studies **		
	B=1.0m	B=2.0m	Johnson (1972)	Germain (1984), Kabbaj (1985)	Madsen and Mei (1969)	Johnson (1972)	Seabra-Santos et al. (1987)
0.67	-	-	2	1	-	-	2 (NB)
0.614	-	-	2	2	-	2 (NB)	-
0.60	-	-	2	2	-	-	2 (NB)
0.55	-	-	3	2	-	3 (NB)	2 (NB)
0.50	2 (NB)	3 (NB)	3	2	3 (NB)	3 (NB)	3 (NB)
0.451	-	-	3	3	-	-	-
0.45	-	-	3	3	-	-	3 (NB)
0.33	3 (NB)	4 (NB)	5	4	-	-	4 (Br)
0.17	4 (NB)	5 (NB)	10	9	-	-	-

Br – breaking waves, NB – nonbreaking waves, * solitons number calculated by the author of this thesis by substituting the considered values of d_r/h into the aforementioned theoretical formulas, ** solitons number provided by previous experimental/numerical studies

3.3 Summary of key results

The key results of the laboratory experiments performed to analyse the influence of a submerged structure of a finite width on the nonlinear transformation of a solitary can be summarized as follows:

- Criterion for a solitary wave breaking over a submerged structure of arbitrary shape, given by Eq. (3.5), is determined for the following range of application: B/L_i=0.2-0.49 and H_i/d_r=0.2-2.37 ($H_i=H_{i,2}$; $L_i=L_{i,2}$). Thereby, the breaking limit supplements the research gap on the breaking of a solitary wave at submerged structures with finite widths, as indicated in Chapter 2.
- The observed breaker types are classified (as the spilling, transition and plunging breakers) and confirmed with those reported in the previous studies.
- Description of the mechanism of the solitary wave fission is provided, including: (i) determination of the number of the generated solitons, (ii) analysis of the spatial evolution of the height of the leading soliton, and (iii) comparison of the observed fission process with that generated over submerged structures of infinite widths.
- Modes of solitary wave evolution are summarized and compared to those reported in the previous investigations.

4. Description, validation and modifications of numerical model COULWAVE

The main features of the numerical model for water wave propagation COULWAVE, selected to perform the numerical investigation on the hydraulic performance of an artificial reef under tsunami-like solitary wave conditions, are described in this chapter, including the equations governing wave motion and the numerical scheme. In order to assure the accuracy of the numerical results and to test the model application, the programme is validated by means of data provided by own experiments (see Chapter 3) and available literature. On the basis of the conclusions of the literature review (see Chapter 2) and the results of the model validation, slight modifications of the source code are introduced.

4.1 Major features of the numerical model

4.1.1 Governing equations

The numerical model COULWAVE was developed by Lynett (2002) to deal with the propagation of water waves over variable bathymetry, including waves generated by submarine landslides, and to predict the corresponding nearshore wave transformation associated with decreasing water depth. The programme features have been continuously improved – a version of the original model for parallel computing has been developed (Sitanggang et al., 2006) and recently, a more stable numerical scheme based on higher-order Finite Volume Method (FVM) has been implemented (Kim et al., 2009).

The numerical code is based on the fully nonlinear, extended Boussinesq-type equations derived by Liu (1994), which allow to model the evolution of highly nonlinear ($\varepsilon=H/h=0(1)$) and dispersive waves. The enhancement of the linear dispersion characteristics was achieved by introducing a concept of a multi-layer approach, according to which the water column can be divided into an arbitrary number of layers with the corresponding velocity profiles matched at the arbitrary specified interfaces (Lynett, 2002). With a one-layer model, the dispersion properties are accurate up to kh~3, what corresponds to the extended Boussinesq-type equations (e.g. by Nwogu, 1993; Liu, 1994 or Wei et al., 1995).

By introducing a two-layer model, improvement of the dispersion properties up to kh~8, with the second-order nonlinear behaviour up to kh~6, was reported (Lynett, 2002).

The horizontal velocity field is evaluated at an arbitrary water level z_α, postulated by Nwogu (1993) to be z_α=-0.531h in order to obtain the best agreement between the resultant and the linear dispersion characteristics.

Under the assumption of inviscid flow, the Boussinesq-type equations in the original form do not account for energy dissipation, which in nature is caused by frictional losses and turbulent effects accompanying wave breaking. This main weakness can be however overcome by introducing ad-hoc dissipative terms in the momentum equation: $\boldsymbol{R}_{br}$ representing the energy dissipation due to wave breaking, $\boldsymbol{R}_{bf}$ due to bottom friction and $\boldsymbol{R}_{sev}$ describing the eddy viscosity dissipation in the Smagorinsky-type subgrid model.

The fully nonlinear, extended Boussinesq-type model for two horizontal (2DH) wave propagation over varying bathymetry, accounting for energy dissipation, consists of the mass (Eq. 4.1) and momentum (Eq. 4.2) conservation equations:

$$\eta_t + \nabla \cdot \left[(\eta + h) \mathbf{u}_\alpha \right] - \nabla \cdot \left\{ \left[\frac{\eta^3 - h^3}{6} - \frac{z_\alpha^2 (\eta + h)}{2} \right] \nabla (\nabla \cdot \mathbf{u}_\alpha) + \left[\frac{\eta^2 - h^2}{2} - z_\alpha (\eta + h) \nabla (\nabla \cdot (h \mathbf{u}_\alpha)) \right] = 0 \tag{4.1}$$

$$\mathbf{u}_{\alpha t}+(\mathbf{u}_\alpha\cdot\nabla)\mathbf{u}_\alpha+g\nabla\eta+\frac{1}{2}z_\alpha^2\nabla(\nabla\cdot\mathbf{u}_{\alpha t})+z_\alpha\nabla\left[\nabla\cdot(h\mathbf{u}_\alpha)_t\right]$$
$$+\left[\nabla\cdot(h\mathbf{u}_\alpha)\right]\nabla\left[\nabla\cdot(h\mathbf{u}_\alpha)\right]-\nabla\left[\eta\left(\nabla\cdot(h\mathbf{u}_\alpha)_t\right)\right]+(\mathbf{u}_\alpha\cdot\nabla z_\alpha)\nabla\left[\nabla\cdot(h\mathbf{u}_\alpha)\right]$$
$$+z_\alpha\nabla\left[\mathbf{u}_\alpha\cdot\nabla\left(\nabla\cdot(h\mathbf{u}_\alpha)\right)\right]+z_\alpha(\mathbf{u}_\alpha\cdot\nabla z_\alpha)\nabla(\nabla\cdot\mathbf{u}_\alpha)+\frac{1}{2}z_\alpha^2\nabla\left[\mathbf{u}_\alpha\cdot\nabla(\nabla\cdot\mathbf{u}_\alpha)\right] \quad (4.2)$$
$$+\nabla\left\{-\frac{1}{2}\eta^2\nabla\cdot\mathbf{u}_{\alpha t}-\eta\mathbf{u}_\alpha\cdot\nabla\left[\nabla\cdot(h\mathbf{u}_\alpha)\right]+\eta\left[\nabla\cdot(h\mathbf{u}_\alpha)\right]\nabla\cdot\mathbf{u}_\alpha\right\}$$
$$+\nabla\left\{\frac{1}{2}\eta^2\left[\nabla\cdot\mathbf{u}_\alpha\right]^2-\mathbf{u}_\alpha\cdot\nabla(\nabla\cdot\mathbf{u}_\alpha)\right\}-\mathbf{R}_{br}+\mathbf{R}_{bf}+\mathbf{R}_{sev}=0$$

where subscript "t" denotes time derivative, g the gravitational acceleration, η the free surface elevation, $\mathbf{u}_\alpha=[u_\alpha, v_\alpha]$ the horizontal velocity vector evaluated at water level $z_\alpha=-0.531h$, h the water depth and $\nabla=[\partial/\partial x,\partial/\partial y]$ the horizontal gradient vector.

a) Wave breaking

The additional terms in the momentum equation, representing the energy dissipation due to the turbulence generated by wave breaking, are expressed in the following form in case of two-horizontal dimensional (2DH) wave propagation:

$$\mathbf{R}_{br}=\left[F_{br},G_{br}\right] \quad (4.3)$$

with the vector components in x- and y-direction, respectively:

$$F_{br}=\frac{1}{h+\eta}\left\{\left[\upsilon((h+\eta)u_\alpha)_x\right]_x+0.5\left[\upsilon((h+\eta)u_\alpha)_y+((h+\eta)v_\alpha)_x\right]_y\right\} \quad (4.4)$$

$$G_{br}=\frac{1}{h+\eta}\left\{\left[\upsilon((h+\eta)v_\alpha)_y\right]_y+0.5\left[\upsilon((h+\eta)v_\alpha)_x+((h+\eta)u_\alpha)_y\right]_x\right\} \quad (4.5)$$

where subscripts "x" and "y" represent spatial partial derivatives.

A modified eddy viscosity concept developed originally by Zelt (1991) is employed to introduce the energy losses generated during wave breaking event. As in the nature, the eddy viscosity υ varies both in space and time, and is located in front of a breaking wave through its dependence on free surface slope η_t:

$$\upsilon=B_r\delta_b(h+\eta)\eta_t \quad (4.6)$$

Mixing length coefficient δ_b, set in the model to 6.5, corresponds to the width of a bore that is generally of the order of several times the local water depth.

Parameter B_r controls the rate of wave energy dissipation and prevents from a sudden generation of breaking event by varying smoothly between the values $B_r=0.0$ (no breaking) and $B_r=1.0$ (fully developed breaking):

$$B_r = \begin{cases} 0 & \eta_t \le 2\eta_t^* \\ \dfrac{\eta_t}{\eta_t^*} - 1 & \eta_t^* < \eta_t < 2\eta_t^* \\ 1 & \eta_t > \eta_t^* \end{cases} \tag{4.7}$$

Unlike the breaking criterion determined by Kennedy et al. (2000) in terms of the shallow water velocity $c=(gh)^{0.5}$, nonlinear wave velocity $c=[g(h+\eta)]^{0.5}$ is employed by Lynett (2002) to define the initiation and cessation of the breaking event. It is assumed that wave breaking is generated when the time derivative of surface elevation η_t exceeds the limit $\eta_t^{(I)}$, and is completed when η_t reaches the value $\eta_t^{(F)}$:

$$\eta_t^* = \begin{cases} \eta_t^{(I)} & t - t_o = 0 \\ \eta_t^{(I)} + \dfrac{t - t_o}{T^*}\left(\eta_t^{(F)} - \eta_t^{(I)}\right) & 0 < t - t_o < T^* \\ \eta_t^{(F)} & t - t_o \ge T^* \end{cases} \tag{4.8}$$

with

$$\eta_t^{(I)} = 0.65\sqrt{g(h+\eta)} \tag{4.9}$$

$$\eta_t^{(F)} = 0.08\sqrt{g(h+\eta)} \tag{4.10}$$

$$T^* = 7\sqrt{(h+\eta)/g} \tag{4.11}$$

Parameter η_t^* represents both initiation and cessation of the breaking event, t is the current time, t_o the time at which wave starts to break, t-to is the age of breaking event and T^* the transition time.

b) Bottom friction

The frictional energy losses are determined according to the quadratic friction law and introduced in the momentum equation in case of 2HD wave propagation in the following vector form:

$$\mathbf{R}_{bf} = \frac{f_b}{h+\eta}\mathbf{u}_b|\mathbf{u}_b| \tag{4.12}$$

where $\boldsymbol{u}_b$ is the horizontal velocity vector evaluated at the sea bottom and f_b is the friction coefficient of the typical range of $f_b=10^{-3}$-10^{-2}.

c) Subgrid turbulent mixing

The effect of the horizontally-distributed eddy viscosity generated by subgrid turbulent processes is required in the models based on the Boussinesq-type equations to resolve properly the flow pattern. For these purposes the Smagorinsky-type subgrid scheme (Smagorinsky, 1963), similar to that by Kirby et al. (1998) is employed, in which the eddy viscosity is dependant on a grid size and derivatives of velocity field (for more details see Kirby et al., 1998).

4.1.2 Numerical scheme

In the original version of the model COULWAVE (Lynett, 2002), a high-order Finite Difference Method is employed to discretize the mass and momentum conservation equations in space, using an un-staggered, uniform grid. The leading-order spatial derivatives are calculated on the basis of five-point difference scheme (with the accuracy of Δx^4), while the three-point centred scheme (of accuracy $\mu\Delta x^2$) is applied to the dispersive terms.

Recently, an optional numerical scheme, based on a higher-order Finite Volume Method (FVM) with shock-capturing, approximate Reimann solver used for leading-order flux terms, has been introduced in order to improve stability and accuracy of the model (Kim et al., 2009). However, approximately 50-100 % of the computational time in comparison to the Finite Difference Method (FDM) is required. The shock-capturing scheme allows to deal with the problem of poorly resolved shocks (i.e. when wave breaking cannot be triggered due to a poor resolution). Similarly to the breaking mechanism commonly used for the fully nonlinear shallow-water equations, the steep front of the breaking wave is frozen and numerical dissipation is added to the system. In this thesis, version of the model based on the FDM is used.

Time-stepping scheme follows the fourth-order corrector-predictor Adams-Bashforth method of the accuracy Δt^4, which consists of the 3rd-order explicit

Adams-Bashforth predictor step and 4th-order implicit Adams-Moulton corrector step. The detailed description of this approach is provided in Strusińska and Oumeraci (2009c) and can be found e.g. in Lynett (2002), Lynett and Liu (2002).

4.1.3 Initial solitary wave profile

The incident solitary wave conditions are modelled according to the analytical solitary wave solution of the weakly nonlinear, extended Nwogu`s (1993) Boussinesq-type equations, derived by Wei and Kirby (1995). Free surface elevation η and horizontal velocity u_α are specified by the following equations:

$$\eta = A_3 \operatorname{sech}^2\left[A_2(x-ct)\right] + A_4 \operatorname{sech}^4\left[A_2(x-ct)\right] \quad (4.13)$$

$$u_\alpha = A_1 \operatorname{sech}^2\left[A_2(x-ct)\right] \quad (4.14)$$

The solitary wave celerity is defined as:

$$c = \sqrt{c' gh} \qquad \left(c' = c/\sqrt{gh}\right) \quad (4.15)$$

and expressed in terms of dimensionless wave speed c', which is approximated from the following formulation:

$$2\alpha\left(c'^2\right)^3 - \left(3\alpha + \frac{1}{3} + 2\alpha\varepsilon\right)\left(c'^2\right)^2 + 2\varepsilon\left(\alpha + \frac{1}{3}\right)\left(c'^2\right) + \alpha + \frac{1}{3} = 0 \quad (4.16)$$

Parameter ε=H/h represents the nonlinear effects. Parameter α is related to the water level at which the horizontal velocity is evaluated (z_α = -0.531h):

$$\alpha = 0.5\beta^2 + \beta \text{ for } \beta = 0.531 \quad (4.17)$$

Parameters A_1, A_2, A_3, A_4 are calculated as follows:

$$A_1 = \frac{c^2 - gh}{c} \quad (4.18)$$

$$A_2 = \left\{\frac{c^2 - gh}{4\left[(\alpha + 1/3)gh^3 - \alpha h^2 c^2\right]}\right\}^{0.5} \quad (4.19)$$

$$A_3 = \frac{c^2 - gh}{3\left[(\alpha + 1/3)gh - \alpha c^2\right]} h \quad (4.20)$$

$$A_4 = -\frac{\left(c^2 - gh\right)^2}{2ghc^2} \frac{\left[(\alpha + 1/3)gh + 2\alpha c^2\right]}{\left[(\alpha + 1/3)gh - \alpha c^2\right]} h \tag{4.21}$$

4.2 Validation of the numerical model

Validation of the numerical model COULWAVE, presented in this chapter, is performed by means of the following data:

- Own laboratory experiments on the nonlinear transformation of a solitary wave over a submerged structure of a finite width, described in Chapter 2. The data is additionally compared with the results obtained by means of the COBRAS model.
- Laboratory experiments by Matsuyama et al. (2007) and Yasuda (2007) on fission and breaking of a tsunami-like sinusoidal wave during propagation over a continental shelf.
- Laboratory experiments by Grüne et al. (2006) on a tsunami-like solitary wave evolution over a continental shelf.
- Numerical results provided by Sue et al. (2005), reproducing laboratory and numerical investigations of Seabra-Santos et al. (1987) on a tsunami-like solitary wave scattering by an infinite submerged step.

Applicability of the source code to predict the hydraulic performance of a submerged structure under tsunami conditions is examined by comparing the numerical results with the available data, with a particular consideration of the accuracy of the implemented wave breaking scheme and the dispersion properties of the governing equations, crucial in case of generation of nonlinear effects.

In this chapter, discussion of the selected cases will be provided; the details of the entire procedure of the model validation can be found in Strusińska and Oumeraci (2009c, d and e).

4.2.1 Model validation using data from own laboratory experiments

Own laboratory experiments on the nonlinear transformation of a solitary wave with a submerged structure of a finite width are performed in the wave flume of LWI. The detailed description of the experimental set-up and the experimental programme is provided in Chapter 3. Two tests with waves of different nonlinear properties are selected in order to examine the fission process for nonbreaking and breaking waves. Since the most favourable conditions for wave scattering are observed for the smallest submergence depth and the widest reef, irrespective of the structure shape, the following tests are selected to perform the model validation:

- Test No. 17: $H_{i,nom}$=0.06 m, h=0.6 m, B=2.0m, h_r=0.5 m, reef slopes 1:2, termed hereafter "smaller-wave test" (see Appendix A, Table A1).
- Test No. 20: $H_{i,nom}$=0.22 m, h=0.6 m, B=2.0m, h_r=0.5 m, reef slopes 1:2, termed hereafter "larger-wave test" (see Appendix A, Table A1).

The arrangement of the wave gauges is shown in Figure 4.1a for the smaller-wave test and in Figure 4.1b for the larger-wave test.

The fully nonlinear, extended Boussinesq-type theory is applied to perform the one-horizontal dimensional (1HD) simulations, in which the horizontal velocity vector is calculated at water level z_α=-0.531h, as postulated by Nwogu (1993). The following spatial and time resolution is used: Δx=0.04 m, Δt=0.05 s for the smaller-wave test and Δx=0.11 m, Δt=0.05 s for the larger-wave test. The incident solitary wave conditions are determined by two following approaches: (i) using time history of free surface elevation measured at wave gauge WG2 located at a distance of 10.0 m from wave maker, (ii) according to the analytical solitary wave solution of the extended Boussinesq-like model by Nwogu (1993), derived by Wei an Kirby (1995) and presented in Section 4.1.3. Bottom friction is modelled for a constant value of friction coefficient f_b=0.002 along the wave flume (i.e. corresponding to the smooth concrete wave flume bottom and smooth structure surface). However, within the rubble-type wave absorber placed at the end of the wave flume, the friction coefficient is changed in order to predict the same nonabsorbed, reflected wave as in the experiments. In order to attenuate the part of the wave generated by the method, in which the time series records are used to specify the incident wave profile, a sponge layer is added at the left boundary of the numerical domain. Therefore, the re-reflection of the wave from the wave maker, which is observed under the laboratory conditions, cannot be reproduced numerically.

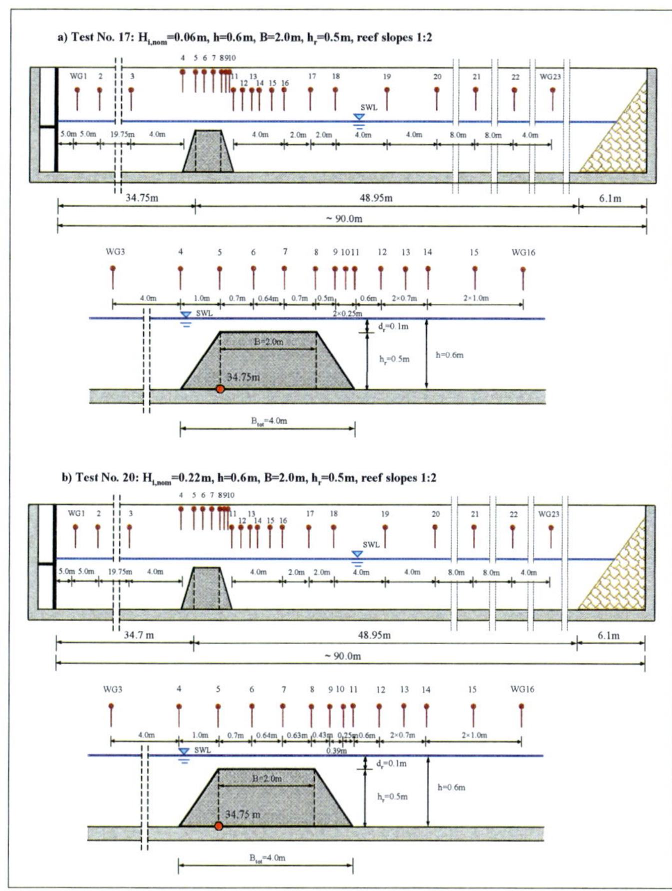

Figure 4.1: Model set-up in own laboratory experiments: a) for smaller-wave test, b) larger-wave test

In order to compare the performance of the source code COULWAVE with a more accurate flow model, the numerical code COBRAS, based on the Reynolds-Averaged-Navier-Stokes equations coupled with k-ε turbulence transport model and Volume of Fluid (VOF) method for tracking the surface elevation, is additionally used. The size of the spatial and time mesh is: Δx=0.06 m, Δz=0.03 m,

Δt=0.05 s for the smaller-wave test and Δx=0.06 m, Δz=0.06 m, Δt=0.05 s for the larger-wave test. The incident solitary wave profile is determined on the basis of solitary wave theory by providing total water depth h and incident wave height H_i. On the left numerical boundary a solid wall representing the flume wall and at the right boundary a porous structure representing the wave absorber in the experiments are defined.

The following conclusions can be made on the basis of the comparative analysis of the measured and the computed time histories of free surface elevation, shown in Figure 4.2 for the smaller-wave test, and in Figure 4.3 for the larger-wave test (for details see Strusińska and Oumeraci, 2009e):

- Generally, the incident wave profile is well predicted in terms of the wave height in comparison to the experiments (see Figures 4.2a and 4.3a). The oscillations at the back part of the wave, intensified with the increasing wave nonlinearity H_i/h, are caused by the imperfections of the wave generation system in the laboratory, and therefore they are not present in the simulations.
- Considering the profile of the wave reflected from the seaward structure slope, a very good agreement in terms of the wave height can be observed (see Figures 4.2a and 4.3a). The computed incident wave height tend to decrease slightly with the propagation distance (what is indicated by the analysis presented in Strusińska and Oumeraci, 2009e), since the resulting generated wave profile represents the solution of the weakly nonlinear extended Boussinesq-type model by Wei and Kirby (1995), not the governing fully nonlinear extended Boussinesq-like equations of Liu (1994). This can explain the phase shift among the reflected wave profiles, since wave reflection occurs later in the numerical simulations than under the laboratory conditions.
- There is generally a good agreement among the profiles of waves propagating over the reef. However, none of the models resolves well the reflection from the landward structure slope during the transition of the wave into the deeper portion of water as well as the evolution of this wave propagating back towards the wave maker (see Figures 4.2b, c and 4.3b, c).

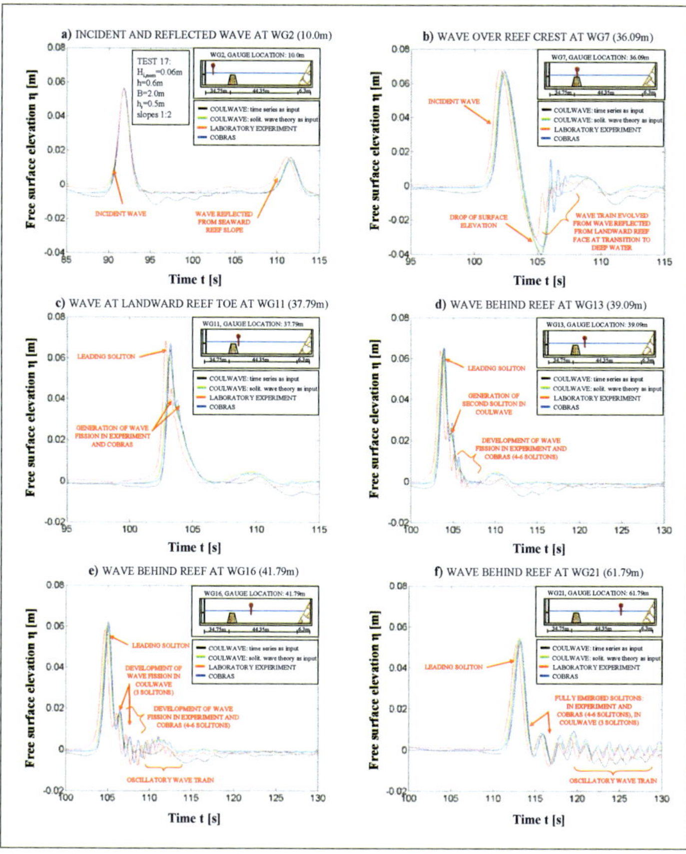

Figure 4.2: Comparison of measured time history of free surface elevation for selected wave gauges in smaller-wave test ($H_{i,nom}$=0.06 m, h=0.6 m, B=2.0 m, h_r=0.5 m, reef slopes 1:2) with those predicted by numerical models COULWAVE and COBRAS

- None of the models could predict accurately the evolution of solitary wave fission. Unlike COULWAVE, in the simulations performed by means of COBRAS, the location of the wave disintegration corresponds very well to that reported in the experiments. For both models the major discrepancies are observed in terms of the soliton amplitudes, with the soliton number better predicted by COBRAS (see Figures 4.2c-f and 4.3d-f). COULWAVE tends to underestimate the number of the generated solitons as wave nonlinearity H_i/h increases. However, this might be due to the coarser grid used for the largest incident waves (due to model instability during wave-structure interaction, simulations with a finer grid cannot be performed).

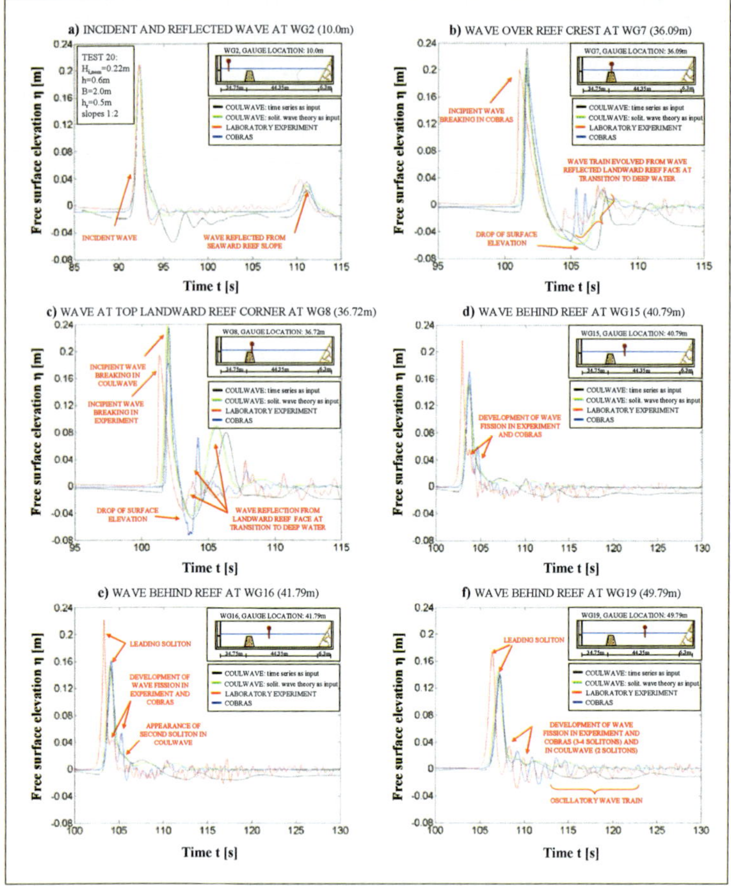

Figure 4.3: Comparison of measured time history of free surface elevation for selected wave gauges in larger-wave test ($H_{i,nom}$=0.22 m, h=0.6 m, B=2.0 m, h_r=0.5 m, reef slopes 1:2) with those predicted by numerical models COULWAVE and COBRAS

- Wave breaking, generated in the larger-wave test, is better reproduced by COBRAS in terms of the height at breaking in comparison to COULWAVE, for which the larger shoaled wave results in a larger breaking wave height (see Figures 4.3b and c).

4.2.2 Model validation using laboratory data by Matsuyama et al. (2007)

In the laboratory experiments of Matsuyama et al. (2007), performed in a Large Wave Flume (Japan) of dimensions 205 m × 2.4 m × 6 m, a tsunami wave was generated as a sinusoidal wave of one cycle only, corresponding to incident wave period T_i=20-120 s and incident wave amplitude a_i=0.005-0.09 m. The wave propagated from a deep-portion of water of constant depth h=4.0 m towards a continental shelf made of sand of varying slope (1:100, 1:150, 1:200), equipped with a frontal slope 1:10 on the one side, and evolving into a beach of a constant slope 1:15 (see Figure 4.4).

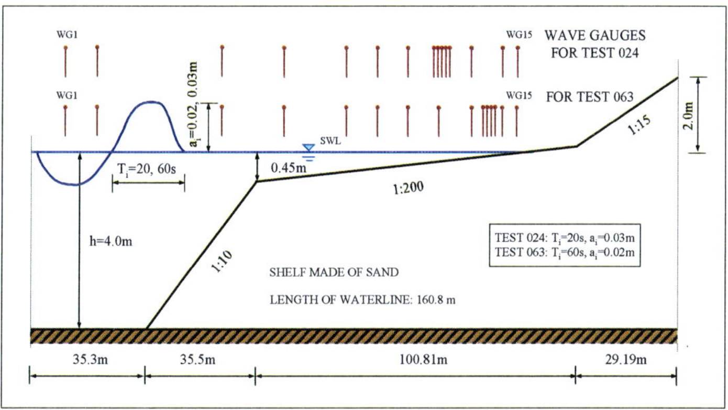

Figure 4.4: Model set-up in experiments of Matsuyama et al. (2007) for shorter-wave test and longer-wave test

Since the mechanism of wave disintegration was reported to be dependent on wave period and shelf slope, the following two cases of different periods and approximately the same amplitude are selected for wave evolution over a shelf of a slope 1:200, corresponding to the most favourable conditions for the fission evolution:

- Test No. 024: T_i=20 s, a_i=0.03 m, shelf slope 1:200, called hereafter "shorter-wave test".
- Test No. 063: T_i=60 s, a_i=0.02 m, shelf slope 1:200, called hereafter "longer-wave test".

It should be remarked that the incident wave resembles a progressive solitary wave, despite the sinusoidal profile of one-cycle only used. Therefore, the terminology related to the solitary wave fission will be herein employed to describe the wave disintegration into a short-period wave train over the shelf.

The 1HD simulations are performed by means of the fully nonlinear, extended Boussinesq-type equations, with the horizontal velocity profile specified according to Nwogu's (1993) approach at water level z_α=-0.531h. The accurate resolution of the tsunami disintegration into a train of shorter-period waves as well as breaking event require fine spatial and time grids: Δx=0.05 m, Δt=0.05 s for the shorter-wave test and Δx=0.06 m, Δt=0.05 s for the longer-wave test. The incident tsunami wave profiles are generated numerically by means of the time history of free water surface recorded at wave gauge WG1, placed at a distance of 10.8 m from wave maker. At the left boundary of the numerical domain, a sponge layer is used to damp both the part of the generated wave travelling towards the left boundary and the wave reflected from the shelf. Frictional wave energy losses are incorporated, with the friction coefficient f_b=0.0025 corresponding to sand bottom.

On the basis of the comparative analysis of the time history of free surface elevation, shown in Figure 4.5 for the shorter-wave test, and in Figure 4.6 for the longer wave test, the following conclusions can be made:

- There is generally a good agreement between the measured and the numerically generated incident "tsunami wave", as shown in Figures 4.5a and 4.6a. However, in case of the longer-wave test, the discrepancy in terms of the incident wavelength can be observed – therefore, the computed wave of a slightly larger period tends to propagate faster than under the laboratory conditions.
- The numerical model predicts very well wave transformation over the shelf in terms of the shoaled wave height (see Figures 4.5b and 4.6b).

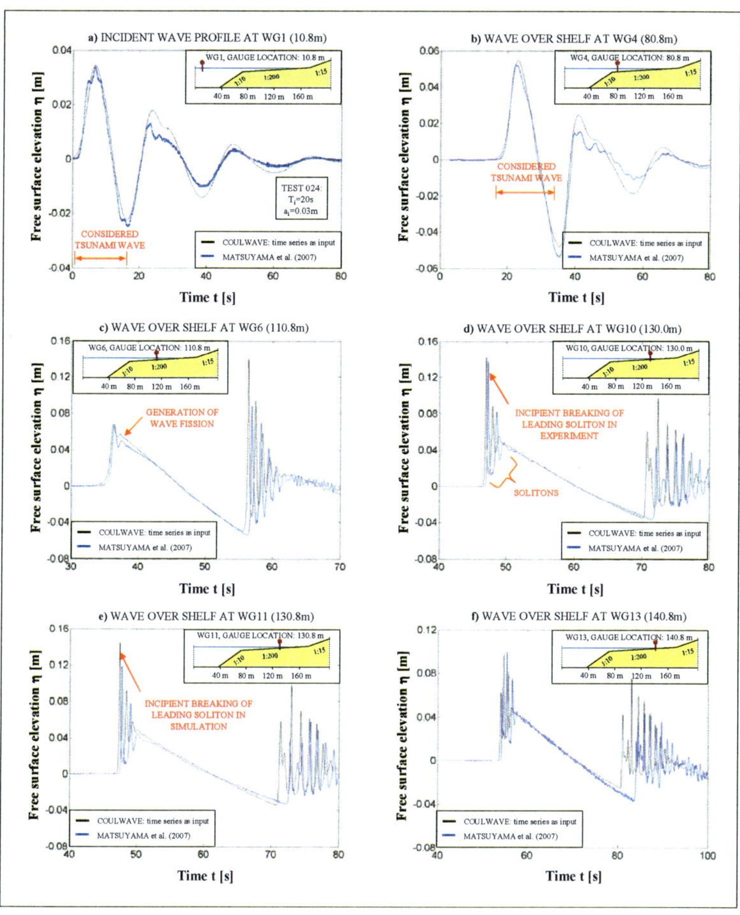

Figure 4.5: Comparison of time history of free surface elevation for selected wave gauges in shorter-wave test (T_i=20 s, a_i=0.03 m) predicted by numerical model COULWAVE and measured by Matsuyama et al. (2007)

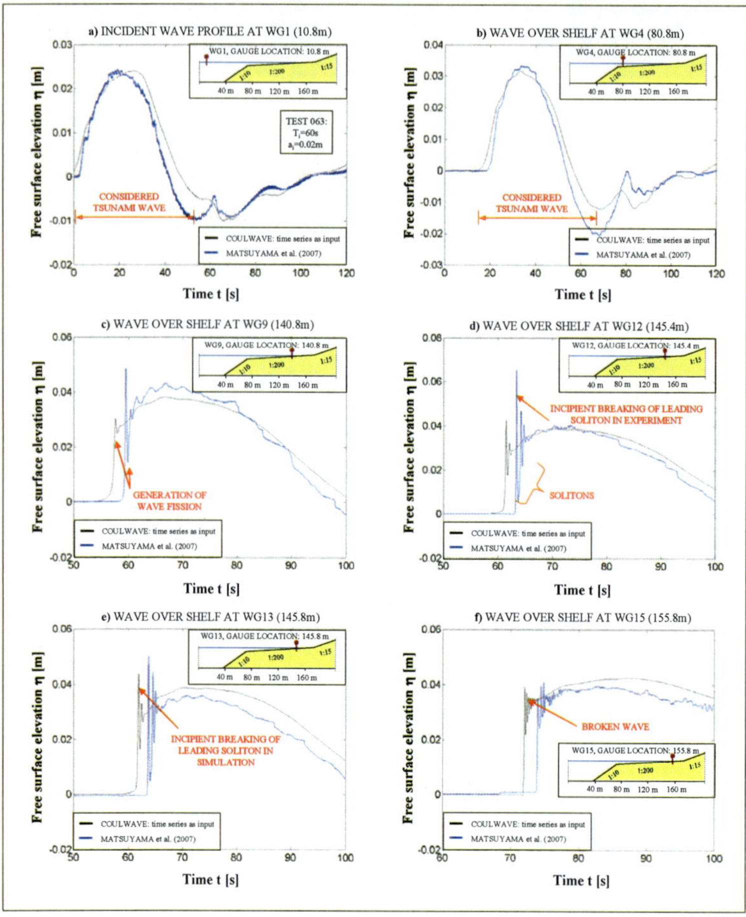

Figure 4.6: Comparison of time history of free surface elevation for selected wave gauges in longer-wave test (T_i=60 s, a_i=0.02 m) predicted by numerical model COULWAVE and measured by Matsuyama et al. (2007)

- Location of the fission initiation in the simulations corresponds very well to that in the experiments. However, there are significant discrepancies related to the number of the emerged solitons and their amplitudes (see Figures 4.5c-f and 4.6c-f). In case of the longer-wave test, numerical instability appearing during the wave generation stage in the simulation does not allow to achieve a better resolution of the generated short-period waves.

- Incipient breaking of the leading soliton, observed for the longer-wave test, occurs slightly earlier in the experiment at wave gauge WG12 at 145.4 m (see Figure 4.6d) than in the simulation at WG13 at 145.8 m (see Figure 4.6e). The discrepancy in the wave height at breaking in the simulation results from the mentioned limitation of the size of the spatial grid used.

4.2.3 Model validation using laboratory data by Grüne et al. (2006)

The transformation of a solitary wave over a continental shelf was investigated laboratory by Grüne et al. (2006) in the Large Wave Flume in Hannover (Germany), which is 300 m long, 5 m wide and 7 m deep. The solitary wave had an incident height H_i=0.1-0.65 m in a constant water depth h=4.5 m. The shelf was made of sand, as a horizontal structure of submergence depth d_r=2.5, equipped with a slope at its seaward side and a beach (slope 1:25) on the other side, as shown in Figure 4.7.

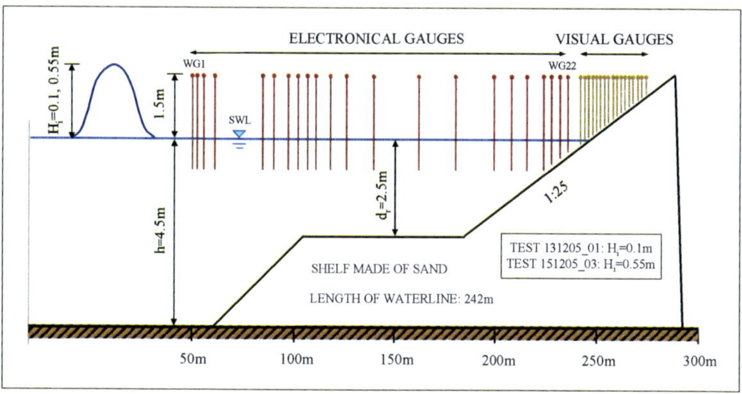

Figure 4.7: Model set-up in experiments of Grüne et al. (2006) for smaller-wave test and larger-wave test

For the purposes of the model validation, the following two cases are selected, representing breaking and nonbreaking tsunami conditions:

- Test No. 131205_01: H_i=0.1 m, termed hereafter "smaller-wave test".
- Test No. 151205_03: Hi=0.55 m, termed hereafter "larger-wave test".

The fully nonlinear, extended Boussinesq-type equations are employed to perform the 1HD simulations, in which the horizontal velocity profile is evaluated at water level z_α=-0.531h, following Nwogu's (1993) theory. The following spatial and time resolution are used for both the smaller- and the larger-wave tests: Δx=0.04 m, Δt=0.05 s. The incident solitary waves are determined in the model by two different methods: (i) by means of the time history of free water surface elevation recorded at wave gauge WG1, located at a distance of 50.1 m from the wave maker, (ii) according to the analytical solitary wave solution provided by Wei and Kennedy (1995). At the left boundary of the numerical domain, a sponge layer is used to damp both the part of the generated wave travelling towards the left boundary and the wave reflected from the shelf. Frictional wave energy losses are incorporated, with the friction coefficient f_b=0.0025 corresponding to the sand bottom.

The comparative analysis of the predicted and measured time history of free surface elevation, presented in Figure 4.8 for the smaller-wave test, and in Figure 4.9 for the larger wave test, indicates as follows:

- Generally, the incident solitary wave profile is well predicted in terms of the wave height and wavelength in comparison to the measured wave (see Figures 4.8a and 4.9a). However, the oscillations of the surface elevation preceding the generated wave and resulting from the imperfections of the wave generation system cannot be reproduced numerically using solitary wave theory as the input data implemented in the numerical models, since the entire wave profile is above still water level.
- There is a satisfactory agreement between the profiles of the wave shoaling over the seaward slope (see Figures 4.8b and 4.9b) as well as the profile of the wave propagating in the constant water depth over the shelf (see Figures 4.8c, d and 4.9c, d). The numerical model tends however to overestimate slightly the height of the progressive wave, particularly during the shoaling over the seaward shelf slope.
- The complex wave reflection, first from the seaward shelf slope and then from the sloping beach, including wave run-down, is well simulated (see

Figures 4.8a-f and 4.9a-f).

- Incipient wave breaking, occurring in the larger-wave test, is very well predicted in terms of the breaking location. The wave height at breaking is however overestimated in comparison to the experiments (see Figure 4.9e).

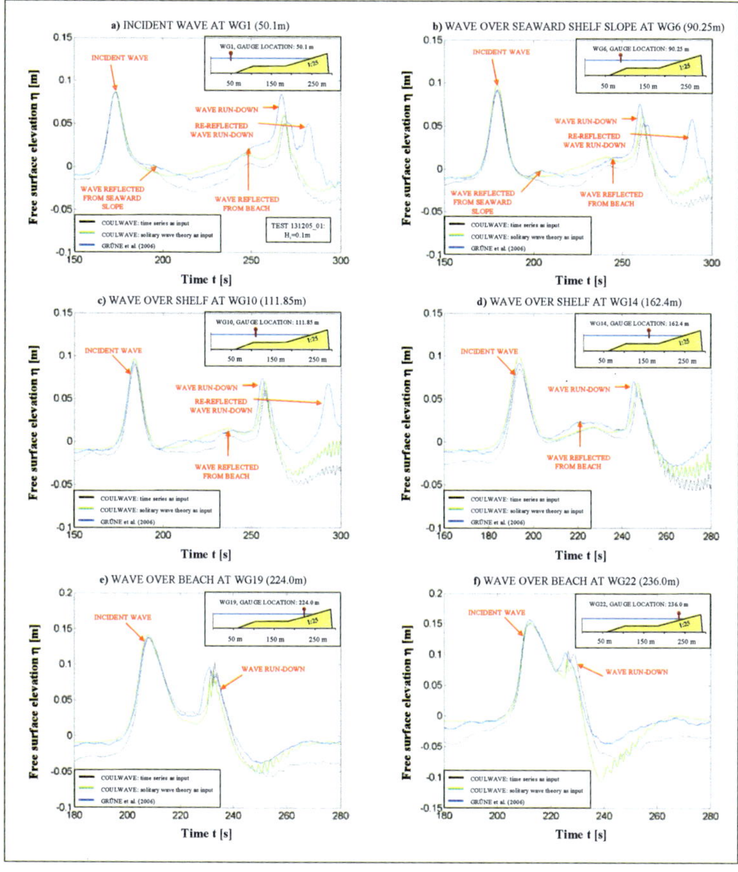

Figure 4.8: Comparison of time history of free surface elevation for selected wave gauges in smaller-wave test (H_i=0.1 m) predicted by numerical model COULWAVE and measured by Grüne et al. (2006)

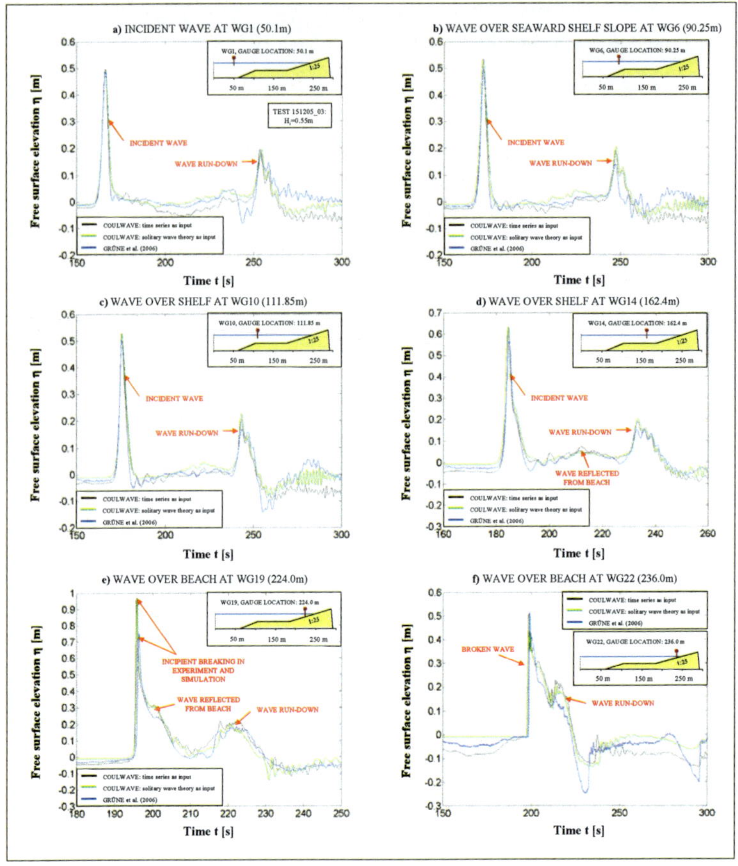

Figure 4.9: Comparison of time history of free surface elevation for selected wave gauges in smaller-wave test (H_i=0.55 m) predicted by numerical model COULWAVE and measured by Grüne et al. (2006)

4.3 Modifications of the numerical model

The commonly employed method for the estimation of the hydraulic performance of a low-crested structure is based on the determination of the so-called energy coefficients: wave transmission K_t, wave reflection K_r and wave energy dissipation coefficients K_d. In case of a regular wave transmission, i.e. with no higher harmonics/solitons induced over barrier crest, the coefficients are expressed in terms of wave height ratios according to Eq. 2.12. If wave transmission over the obstacle is accompanied by the generation of higher harmonics/solitons, it is recommended to calculate the energy coefficients on the basis of wave energy ratios, according to Eq. 2.14 (e.g. Chang et al., 2001; Lin, 2004).

Since the computed unknowns in the Boussinesq-type equations are the free surface elevation and the horizontal velocity vector, it is necessary to implement a procedure for the calculation of solitary wave energy in the numerical source code. Total wave energy E_{tot} of a progressive wave consists of potential E_p and kinetic energy E_k:

$$E_{tot} = E_p + E_k \tag{4.22}$$

The derivation procedure of the total wave energy components is provided below.

4.3.1 Potential wave energy

Potential wave energy results from the displacement of water mass from its equilibrium against the gravitational field.

There are two numerical sub-domains to be distinguished in the source code, as presented in Figure 4.10. In the first region, extended up to a shoreline, wave propagation is governed by the fully nonlinear, extended Boussinesq-type equations. In the second region with "negative values of water depth", wave run-up occurs and is predicted by the fully nonlinear shallow water wave theory. This requires definition of the potential energy for these two sub-domains, separately.

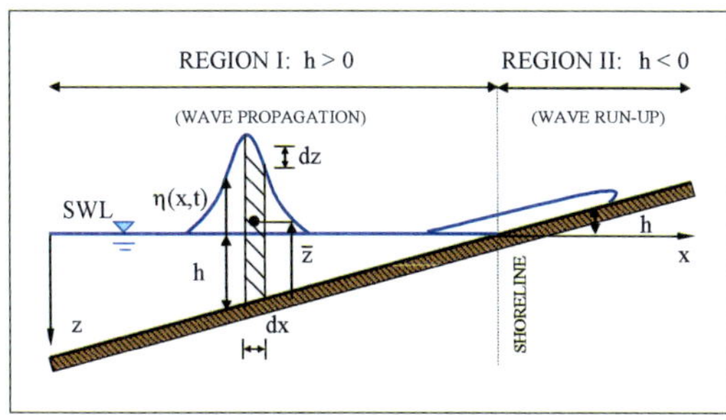

Figure 4.10: Definition sketch of potential wave energy

a) Potential wave energy per unit area in Region I

The potential energy per unit area attached to the wave only can be calculated as follows (Longuet-Higgins, 1974; Liu and Cheng, 2001):

$$E_p = 0.5\rho g\eta^2 \tag{4.23}$$

b) Potential wave energy per unit area in Region II

In the second region, the small water column is stretched within the profile of the wave that undergoes the process of wave run-up. Therefore, the energy per unit area attached to the wave can be determined directly as:

$$E_p = 0.5\rho g(\eta + h)^2 \tag{4.24}$$

Since the free surface elevation η is defined in the source code in respect to the still water level (SWL), the actual water displacement in the Region II corresponds to the computed surface elevation η reduced by the land elevation h (taking negative values in this sub-domain).

4.3.2 Kinetic wave energy

The kinetic energy is related to the movement of water particles both in horizontal and vertical directions. The energy per unit area contained in a small water column yields:

$$dE_k = 0.5dm \cdot \mathbf{U}^2 \tag{4.25}$$

with differential mass dm per unit area given in the following form:

$$dm = \rho dz \tag{4.26}$$

Velocity vector $\boldsymbol{U}$ can be represented by the horizontal velocity vector $\boldsymbol{u}$ and vertical velocity w:

$$\mathbf{U} = [\mathbf{u}, w], \qquad \mathbf{u} = [u, v] \tag{4.27}$$

however for 1HD wave propagation v=0.

The kinetic energy per unit area can be obtained by integrating dE_k over the total water depth:

$$E_k = 0.5\rho \int_{-h}^{\eta} \mathbf{U}^2 dz \tag{4.28}$$

a) Kinetic wave energy per unit area in Region I

The horizontal and vertical velocity profiles for waves generated by underwater landslides are derived by Lynett and Liu (2002) for the fully nonlinear, extended Boussinesq-type equations. In terms of dimensionless variables they read as follows:

$$\mathbf{u} = \mathbf{u}_\alpha - \mu^2 \left\{ 0.5\left(z^2 - z_\alpha^2\right)\nabla\left(\nabla \cdot \mathbf{u}_\alpha\right) + \left(z - z_\alpha\right)\nabla\left[\nabla \cdot \left(h\mathbf{u}_\alpha\right) + \frac{h_t}{\varepsilon}\right]\right\} \tag{4.29}$$

$$w = -z\left(\nabla \cdot \mathbf{u}_\alpha\right) - \nabla \cdot \left(h\mathbf{u}_\alpha\right) - \frac{h_t}{\varepsilon} \tag{4.30}$$

with $\boldsymbol{u}_\alpha$ denoting the horizontal velocity vector evaluated according to Nwogu's (1993) approach at water level z_α=-0.531h. Wave nonlinearity ε and frequency dispersion parameter μ are given by Eqs. (2.2) and (2.3), respectively. Neglecting wave generation by underwater landslide (i.e. $h_t/\varepsilon=0$), and assuming 1HD wave propagation, the velocity profiles can be simplified to the following form written in terms of dimensional variables:

$$u = u_\alpha - \left[0.5\left(z^2 - z_\alpha^2\right) u_{\alpha xx} + \left(z - z_\alpha\right)\left(hu_\alpha\right)_{xx}\right] \tag{4.31}$$

$$w = -zu_{\alpha x} - \left(hu_\alpha\right)_x \tag{4.32}$$

The total kinetic wave energy per unit area can be found by inserting Eqs. (4.31) and (4.32) into Eq. (4.28):

$$E_k = 0.5\rho\int_{-h}^{\eta}\left(u^2 + w^2\right)dz =$$
$$0.5\rho\Big\{u_\alpha^2(\eta+h)+$$
$$u_\alpha\Big\{u_{\alpha xx}\left[-1/3\left(\eta^3+h^3\right)+z_\alpha^2(\eta+h)\right]+(hu_\alpha)_{xx}\left[-\left(\eta^2-h^2\right)+2z_\alpha(\eta+h)\right]\Big\}+$$
$$u_{\alpha xx}^2\left[1/20\left(\eta^5+h^5\right)-1/6z_\alpha^2\left(\eta^3+h^3\right)+1/4z_\alpha^4(\eta+h)\right]+$$
$$u_{\alpha xx}(hu_\alpha)_{xx}\left[1/4\left(\eta^4-h^4\right)-1/3z_\alpha\left(\eta^3+h^3\right)-1/2z_\alpha^2\left(\eta^2-h^2\right)+z_\alpha^3(\eta+h)\right]+$$
$$(hu_\alpha)_{\alpha xx}^2\left[1/3\left(\eta^3+h^3\right)-z_\alpha\left(\eta^2-h^2\right)+z_\alpha^2(\eta+h)\right]+$$
$$1/3u_{\alpha x}^2\left(\eta^3+h^3\right)+u_{\alpha x}(hu_\alpha)_x\left(\eta^2-h^2\right)+(hu_\alpha)_x^2(\eta+h)\Big\}$$

(4.33)

b) Kinetic wave energy per unit area in Region II

Under the assumptions of the fully nonlinear shallow water wave theory, governing the process of wave run-up in the Region II, the vertical velocity *w* can be neglected, while the horizontal velocity profile becomes constant over water depth:

$$\mathbf{U} = [\bar{\mathbf{u}}, w], \qquad \bar{\mathbf{u}} = [\bar{u}, \bar{v}], \qquad w = 0, \tag{4.34}$$

with $\bar{v}$=0 for 1HD wave propagation.

Horizontal velocity vector $\mathbf{u}_\alpha$ can represent in this case the depth-averaged velocity vector $\bar{\mathbf{u}}$:

$$\bar{\mathbf{u}} = \mathbf{u}_\alpha = [u_\alpha, v_\alpha] \tag{4.35}$$

The kinetic wave energy per unit area and one-horizontal dimensional (1HD) wave propagation yields:

$$E_k = 0.5\rho u_\alpha^2(\eta + h) \tag{4.36}$$

4.3.3 Verification of wave energy calculations

In order to control if wave energy, calculated according to the equations derived above, is conserved, a numerical simulation is performed in which no energy losses occur (i.e. wave breaking and bottom friction are neglected). The numerical set-up is shown in Fig. 4.11. A solitary wave of the incident height H_i=0.1 m propagates from water depth h=1.0 m towards a continental shelf of height h_r=0.5 m and infinite width (B→∞). The shelf consists of a seaward slope 1:20 and a horizontal part of submergence depth d_r=0.5 m.

Reflective wall boundaries are employed at the beginning and the end of the numerical domain, which is discretized in space using the resolution Δx=0.05 m and in time using the grid size Δt=0.05 s. The one horizontal dimensional simulation is performed using the fully nonlinear, extended Boussinesq-type of equations, in which the Nwogu`s (1993) concept of calculation of the velocity field at water depth z_α=-0.531h is included.

Changes of wave energy in time as the wave propagated from the deep-water part towards the shelf and along the shelf are presented in Fig. 4.12.

The kinetic and potential energy per unit area computed by COULWAVE is integrated over the entire numerical domain, normalized by the value of the total wave energy recorded at time step t=5.0 s and plotted for each time step (for 70 s of wave propagation, during which the wave reaches the end of the shelf):

$$E_p = \int_0^{endx} 0.5\rho g\eta^2 dx \tag{4.37}$$

$$E_k = \int_0^{endx}\int_{-h}^{\eta} 0.5\rho u_\alpha^2 \, dx \, dz \tag{4.38}$$

The total energy is determined as a sum of the integrated potential and kinetic energy.

The results indicate conservation of wave energy in the entire numerical domain during wave propagation over the horizontal sea bed, wave transformation due to shoaling over the shelf slope (where part of the kinetic energy is transferred into potential energy) and finally over the horizontal shelf, where process of solitary wave disintegration into solitons is observed. It is also noticeable that for weakly nonlinear solitary waves (here H_i/h=0.1), potential and kinetic energy are approximately equal (as for the linear wave theory), as indicated by Longuet-Higgins (1974).

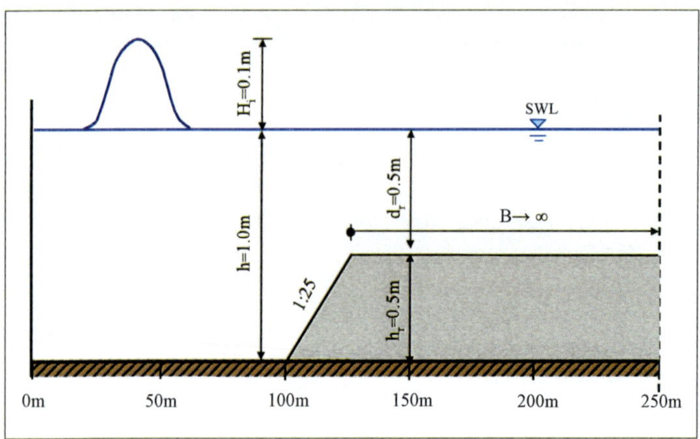

Figure 4.11: Numerical set-up for verification of wave energy calculations

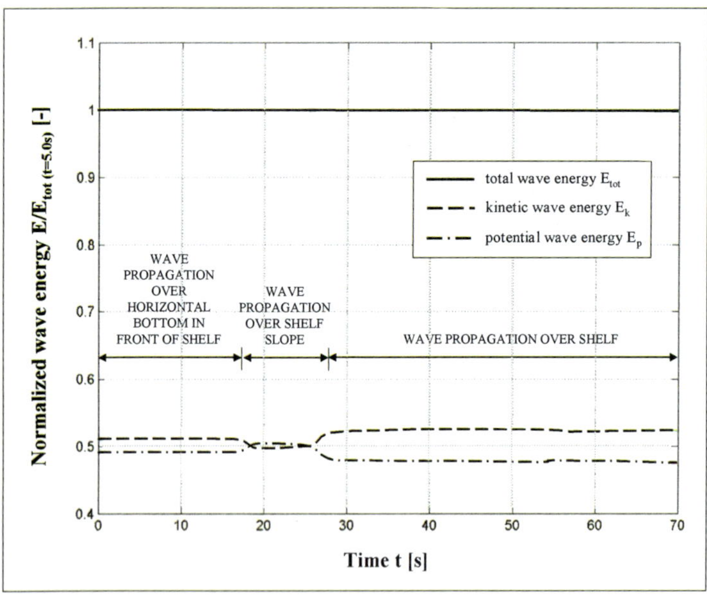

Figure 4.12: Time history of potential, kinetic and total wave energy

4.4 Summary of key results

The key results of the validation of the numerical model COULWAVE and modifications of the source code can be summarized as follows:

- Comparison of the computed results with the data available from own laboratory experiments (see Chapter 3) and from the previous studies (Matsuyama et al., 2007; Yasuda et al., 2007; Grüne et al., 2006; Sue et al. (2005) together with Seabra-Santos et al., 1987) indicate a relatively good overall performance of the model, particularly when predicting nonlinear wave transformation over submerged features (inception of wave breaking, generation/evolution of fission process).
- The numerical scheme and breaking sub-model, originally implemented in COULWAVE, remains unchanged. Calculations of potential and kinetic energy (per unit area) of a progressive solitary wave are introduced in the original source code in order to determine the hydraulic performance of artificial reef as a function of wave transmission, wave reflection and wave energy dissipation (according to Eq. 2.14).

In order to improve model stability, to increase the accuracy of the computed results and also to reduce the computational costs, the implementation of a non-uniform spatial grid would be required, with a coarser resolution used in the farfield and a finer grid used in the nearfield (in this case around and behind an artificial reef).

5. Numerical analysis of hydraulic performance of artificial reef under tsunami conditions

The analysis of the results of the numerical investigation on the hydraulic functioning of an artificial reef under tsunami-like solitary wave conditions, performed by means of the Boussinesq-type model COULWAVE, is provided in this chapter. The primarily goal of the analysis is to determine the effectiveness of the submerged structure in terms of wave transmission, wave reflection and wave energy dissipation for different incident wave characteristics (relative wavelength L_i/h and relative wave height H_i/h) and various reef configurations (relative structure width B/L_i and relative submergence depth d_r/h). The numerically predicted evolution modes of the solitary wave are additionally summarized and compared to those observed in the experiments performed in the twin-wave flume of LWI (see Chapter 3), including a brief description of the wave breaking conditions over the reef and the generation of wave fission process. Such an analysis performed on a basis of experimental results is nowadays still inappropriate due to the much shorter wavelength generated as compared to that obtained from numerical simulations, as mentioned in the introductory chapter.

5.1 Numerical set-up and procedure

5.1.1 Bathymetry in front of and behind the reef

The implemented sea bottom profile represents a simplified case of possible bathymetry configurations that can be found in coastal regions. In order to simulate a tsunami more realistically (e.g. Grüne et al., 2006; Lynett, 2007), the incident wave is located in the farfield at water depth h_o, from where it propagates towards a shelf and an artificial reef placed on it (Figure 5.1a). The same bathymetry, however without the reef, was also used to evaluate the incident wave conditions and the planned reef location with depth h, according to Figure 5.1b.

In the numerical simulations, the shelf slope 1:50 is selected. The length of the sloping part of the sea bed x_1 is limited by water depth h_o (at which the initial wave profile is placed) and depth h, at which the artificial reef is located (see Figure 5.1).

Variation of the offshore water depth ho depends on the predetermined local incident wave conditions (see Table 5.2). The water depth at the artificial reef h is kept constant and can be determined arbitrarily (considering the water depth limitations due to the feasibility of the reef construction). Since the bathymetry, the incident wave conditions in the far- and nearfield and the reef geometry are dependant on this water depth, the computed results (performed for h=1.0 m) can be applied by a simple scaling to the water depth at which the reef is constructed. The choice of the water depth h in the numerical investigations is additionally appropriate for the further analysis of the results, performed in terms of dimensionless variables.

Adopting a horizontal sea bed in front of and behind the reef, resembling a horizontal shelf with a frontal slope (see Figure 5.1), instead of a prolongation of the sloping bottom has the advantage of modelling arbitrary reef configurations (i.e. long-crested reefs). The horizontal section in front of the reef x_2, which equals one incident wavelength L_i, was introduced to determine the incident wave conditions (relative wave height H_i/h and relative wavelength L_i/h) more accurately (where H_i and L_i are the height and the length of the local incident wave according to Figure 5.1). Distance x_3 represents the total width of the considered barrier and varies according to the actual relative width of the reef crest B/L_i and relative submergence depth d_r/h. The length of the section x_4 behind the reef ($x_4=7L_i$) results from the distance required to determine wave transmission as well as to ensure that the propagation distance is long enough for the evolution of the wave fission process.

The initial solitary wave is located at a large distance in the farfield on a horizontal part of the bottom that precedes the shelf slope (see Figure 5.1). This distance is required for a generation of the initial wave profile only and is not considered as a part of the numerical domain (i.e. the output data is saved only for the spatial nodes between the toe of the shelf slope and the end of the shelf).

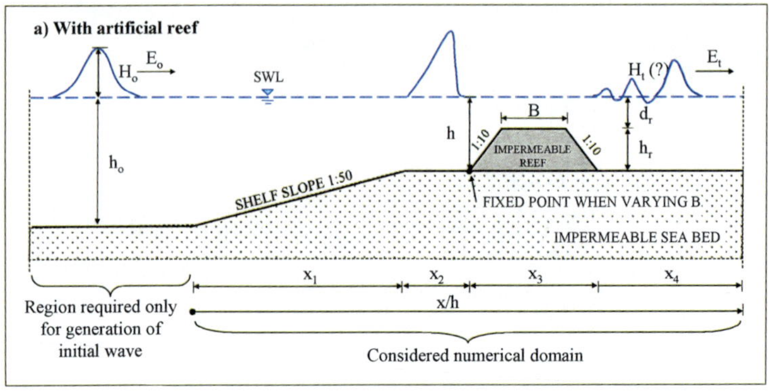

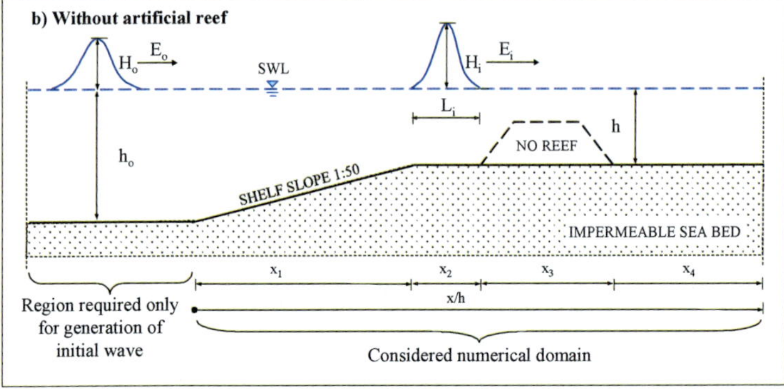

Figure 5.1: Numerical set-up in case of: a) simulations with artificial reef, b) without artificial reef

5.1.2 Reef geometry

The artificial reef investigated in this study is assumed to be impermeable and trapezoidal. The proposed reef geometry, as shown in Figure 1.2, can only be specified in relation to the local incident tsunami wave characteristics and the local water depth. As already mentioned, numerous experimental and numerical investigations performed for various low-crested structures for storm waves represent the only source of information on the design of submerged long-crested structures against tsunami. The conclusions from the present state of knowledge (see Chapter 2) and the results of own laboratory experiments (see Chapter 3) indicate that the parameters playing the most significant role for the wave attenuation performance are relative submergence depth d_r/h, relative reef width B/L_i and relative incident wave height H_i/h.

The maximum submergence depth d_r, which can still contribute meaningfully to wave damping, is, according to the available studies, half of local water depth h at which the submerged structure is constructed (i.e. $d_r/h \leq 0.5$). Due to the limitations of the numerical model COULWAVE that are discussed further in Section 5.1.4, the following range of the relative submergence depth is used in this study: d_r/h=0.5, 0.4 and 0.3. Moreover, numerical tests without the reef are also performed (i.e. d_r/h=1.0, h_r/h=0.0). The relative reef height h_r/h can be obtained from the following relation (see Figure 5.1):

$$h_r = h - d_r \tag{5.1}$$

and equals h_r/h=0.0, 0.5, 0.6, 0.7, respectively.

In order to achieve favourable prematurely wave breaking conditions over the reef, the relation between the width of the structure crown B and the local incident wavelength L_i should be additionally considered (relative structure width B/L_i), as indicated by own laboratory experiments (see Strusińska and Oumeraci, 2009e). As a consequence, the wider the submerged structure in comparison to the incident wavelength, the greater the probability of inception of wave breaking, and the greater the energy losses. However, many authors agree that the optimum width of the structure is conditioned by forcing the incoming wave to break and its further extension does not affect the rate of the energy dissipation substantially (e.g. Dattatri et al., 1978). Nevertheless, very large dimensions of the reef structure will be required to ensure an efficient protection against tsunami. The following range of the relative reef width is selected for this study: B/L_i=0.0-1.0 (with an increment of 0.1), based on the experience made for low-crested structures designed for storm waves.

The steepness of the reef slopes depends on the construction material. Geotextile mega-containers as a protective cover of a reef core made of sand are proposed (see Figure 1.2), for which the typical slope is 1:2 or 1:3. In this study however, reef steepness of 1:10 for both sea- and shoreward slopes is used (see Figure 5.1a), which is more favourable for generation of wave breaking.

Since numerous reef configurations are examined, there is a necessity to determine a generic method of defining the structure geometry in order to obtain identical incident wave conditions for reefs of different widths and submergence depths. For this purpose, the seaward reef toe was selected as a fixed point from which the widening of the structure proceeds shoreward.

Values of all the parameters defining the reef geometry are summarized in Table 5.1.

Table 5.1: Reef geometry used in numerical investigations (see also Figure 5.1)

PARAMETER	VALUE	UNIT
h	1.0	[m]
d_r/h	0.3, 0.4, 0.5 and 1.0 (no reef)	[-]
h_r/h	0.7, 0.6, 0.5 and 0.0 (no reef)	[-]
B/L_i	0.1 – 1.0 (with increment 0.1) and 0.0 (no reef)	[-]
reef slopes	1:10	[-]

5.1.3 Incident wave conditions

In this study, a tsunami wave is reproduced numerically as a solitary wave, since it is the most common wave type used both in laboratory and numerical investigations to reflect the tsunami properties (e.g. Goring, 1979; Grilli et al., 1994; Liu and Cheng, 2001; Grüne et al., 2006; Lynett, 2007). Since the solitary wave profile is specified by the height of the wave and the water depth at which it propagates, changing both parameters would allow to adjust the solitary wavelength (theoretically infinite) to the real tsunami conditions (i.e. to reproduce a tsunami with a range of period of T=2-60 min). In fact, the major shortcoming of the existing studies in tsunami science is the "short-period" characteristic of the tsunami-like solitary waves that are generated under constant water depth conditions (e.g. Goring, 1979; Seabra-Santos et al., 1987; Lin, 2004; Madsen et al., 2008). For this purpose, the concept of a tsunami generation in a deep portion of water (e.g. Grüne et al., 2006; Lynett, 2007) is used in this study to simulate the long-period nature of a tsunami wave more realistically.

For this purpose, two different wave conditions need to be distinguished: *the initial wave characteristic* in the farfield (relative initial wave height H_o/h_o and relative initial water depth h_o/h) and *the local incident wave parameters* (relative incident wave height H_i/h, relative wavelength L_i/h, and local water depth h), specified at the intended reef location in the simulations with no reef structure used (i.e. B/L_i=0.0, d_r/h=1.0), as illustrated in Figure 5.1b. The initial wave profiles in the farfield result from the predetermined local incident wave conditions.

The incident waves in the farfield are generated with relative initial wave heights H_o/h_o=0.021, 0.026, 0.029, 0.03 at relative water depths h_o/h=3.5, 4.0, 4.5, 5.0, respectively. These values are determined by the method of trial and errors and are selected in such a way that the resultant incident wave profile matches the required local incident wave parameters.

The relative local incident wave height, measured as the wave approaches the point corresponding to the planned location of the seaward reef toe, is in the following range: H_i/h=0.10, 0.15, 0.20, 0.25.

Since tsunami properties correspond to shallow-water depth conditions (i.e. $\mu=h/L<0.05$), a relative tsunami wavelength of L_i/h=100 (i.e. μ=0.01) was selected for this study for all considered relative incident wave heights H_i/h:, taking into account the computational effort required for simulating longer waves.

Both initial wave conditions in the farfield and local incident wave characteristics, which are applied in the numerical simulations, are summarized in Table 5.2. Comparison of incident conditions of solitary wave and reef geometry considered in own laboratory experiments and numerical simulations is provided in Table E1 in Appendix E.

Table 5.2: Wave characteristic used in numerical investigations

INITIAL WAVE CONDITIONS IN FARFIELD		LOCAL INCIDENT WAVE CONDITIONS* (determined for B/L_i=0.0, d_r/h=1.0, i.e. no reef)		
H_o/h_o [-]	h_o/h [-]	H_i/h [-]	L_i/h [-]	h [m]
0.021	3.5	0.10	100	1.0
0.026	4.0	0.15	100	1.0
0.029	4.5	0.20	100	1.0
0.030	5.0	0.25	100	1.0

(*) at the planned location of the reef toe and without the reef

5.1.4 Selected numerical parameters and model limitations

The numerical analysis is performed using the fully nonlinear, extended Boussinesq- type of equations, which are implemented in the numerical model

COULWAVE, assuming one-horizontal dimensional wave propagation and one-layer model. The horizontal velocity field is evaluated at water level z_α=-0.531h, according to Nwogu (1993).

A Finite Difference Method (FDM) is used. The spatial grid resolution is expressed as a function of the local water depth *h* and equals $\Delta x/h$=0.1, 0.1, 0.15, 0.15 for the corresponding relative incident wave heights H_i/h=0.1, 0.15, 0.2, 0.25. The time step is kept constant in each performed simulation: Δt=0.05s.

Reflective solid walls are used as a boundary at both ends of the numerical domain in order to prevent from other sources of wave energy damping (occurring when using sponge layers).

Frictional energy losses are neglected in the simulations indicated by preliminary analysis performed for a friction coefficient typical for sand bottom (f_b=0.0025) that was constant in the numerical domain. This allows an easier control of the energy conservation in the run simulations. The submodel of wave breaking, as presented in Chapter 4, is used in an unchanged form.

In the process of the tentative analysis performed for determining the incident wave characteristics and for controlling the energy conservation in the numerical domain, the following limitations of the source code COULWAVE were recognized:

- The implementation of a nonunifrom spatial grid is required to model the process of fission more accurately in case of higher solitary waves – with a coarser grid used in the farfield and a finer one in the vicinity of the structure. In fact, numerical instability appeared in the initial wave profile when applying a finer grid for the cases of $H_i/h>0.25$. Since the implementation of the nonuniform spatial grid scheme was not possible within the time frame of this study, the investigation was limited to the cases with the relative incident wave height in the range of $H_i/h \leq 0.25$.
- Numerical instability appeared for smaller ratios of the relative submergence depth ($0.0<d_r/h<0.3$) as well as steeper shoreward slopes of the reef, as the solitons, generated due to the fission process over the structure, entered the deeper portion of the water behind the reef (see Figure 5.1a). Considering the restriction of wave energy conservation, the use of numerical parameters improving numerical stability by turning on the Smagorinsky (1963) type

scheme for subgrid eddy viscosity, was not possible. Continuous dissipation of wave energy over the entire wave propagation distance was computed when including the subgrid eddy viscosity submodel in the tentative simulations. This resulted in overestimated wave attenuation and thus in unrealistic prediction of the damping performance of an artificial reef. Therefore, the following range of relative submergence depth d_r/h can be examined in this study: $d_r/h=0.3$, 0.4, 0.5 and 1.0 (no reef).

5.1.5 Adopted procedure

A simplified flow chart for the performance of the numerical investigations is shown in Figure 5.2. The scheme is elaborated assuming arbitrary local water depth h (in the simulations assumed to be h=1.0 m) and one relative incident wavelength $L_i/h=100$ for all the examined incident wave heights ($H_i/h=0.10\text{-}0.25$). For each value of relative local incident wave height H_i/h, different relative widths of reef crown ($B/L_i=0.0\text{-}1.0$) are investigated (see Figure 5.2 elaborated exemplarily for $H_i/h=0.20$). Finally, each variation of the relative submergence depth ($d_r/h=0.3\text{-}0.5$, 1.0) is analysed together with the variation of the ratio B/L_i.

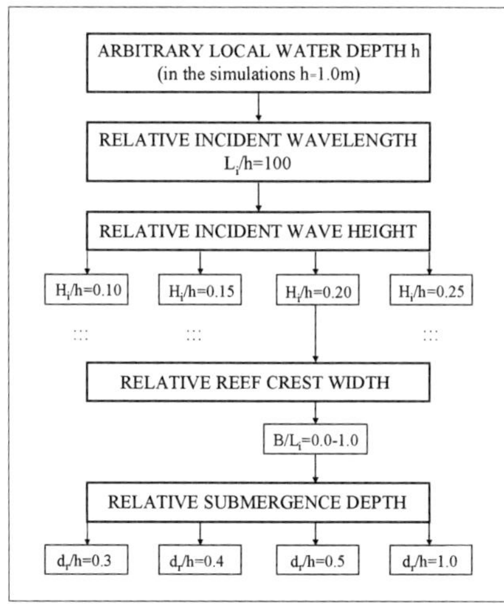

Figure 5.2: Procedure and selected parameters for the performance of numerical investigations

5.2 Calculation of hydraulic performance of reef structure

5.2.1 Definition of energy coefficients K_t, K_r and K_d

The hydraulic performance of artificial reef under tsunami conditions is determined on basis of wave energy coefficients (wave transmission K_t, wave reflection K_r and wave energy dissipation K_d). These coefficients are expressed in terms of wave energy according to Eqs. 2.14 and 2.15, since tsunami fission over the horizontal shelf and over the artificial reef is expected, as indicated by the laboratory experiments (see Chapter 3).

In order to determine energy coefficients K_t, K_r and K_d, solitary wave energy per unit area is first calculated according to the derived equations, implemented in COULWAVE (see Eqs. 4.23 and 4.24 for potential energy E_p and Eqs. 4.33 and 4.36 for kinetic energy E_k). Potential and kinetic energy per unit area is computed for each spatial node of the considered numerical domain at each time step. This output data is then used by a procedure developed by means of Matlab software to calculate potential, kinetic and total wave energy per unit width for the considered numerical domain at each time step of the simulation by means of trapezoidal numerical integration scheme. The total wave energy E_{tot} is calculated as a sum of the potential and kinetic energy.

The method of determining the incident, transmitted, reflected and dissipated wave energy is based on the analysis of time history of kinetic, potential and total wave energy per unit width, computed for the entire numerical domain and plotted for each time step of the simulations. The considered numerical domain, as already explained in Section 5.1.1, is stretched from the toe of the shelf slope to the end of the shelf (see Figure 5.1). Details of the procedure applied to determine the incident, transmitted, reflected and dissipated wave energy are provided in the next sections.

a) Definition of incident wave energy in case of no reef and nonbreaking waves

The incident wave energy per unit width is defined for the case of B/L_i=0.0 and d_r/h=1.0 (no reef) as the difference between the energy of the wave entering the horizontal shelf and the energy of the wave reflected from the shelf slope and leaving the numerical domain through the left boundary (see Figure 5.4). Kinetic, potential and total energy are normalized by the value of the total wave energy computed at normalized time $t(g/h)^{0.5}$=344.53, while time is normalized by $t(g/h)^{0.5}$. Time $t(g/h)^{0.5}$=0 corresponds to the entrance of a solitary wave into the numerical domain (i.e. the wave starts propagating over the shelf slope).

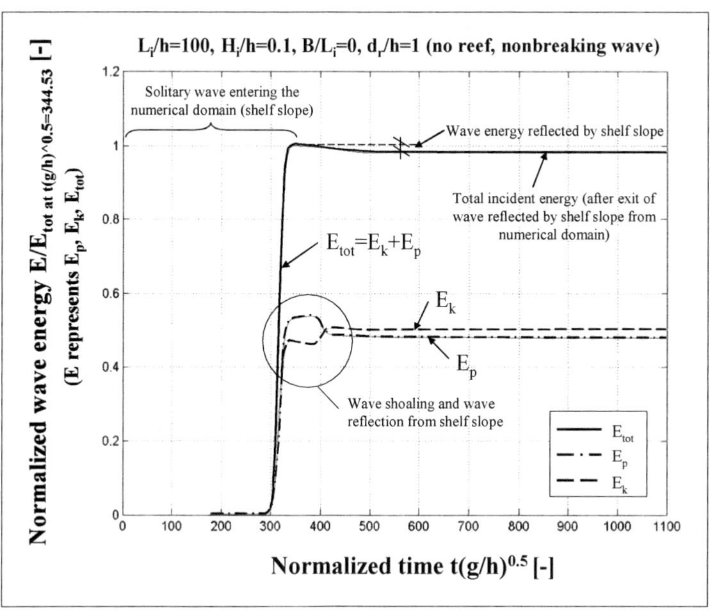

Figure 5.4: Determination of incident wave energy in case of no reef and nonbreaking waves

The initial growth of the energy corresponds to the entrance of the solitary wave into the numerical domain. The total energy is subsequently reduced by the part of the wave energy that is reflected from the shelf slope. Once the reflected wave exits the numerical domain, the total energy remains constant as the wave propagates over the horizontal shelf. Potential energy increases as the wave shoals over the shelf slope, while kinetic energy decreases. This transfer of kinetic energy into potential energy is also induced by a partial reflection of the incident wave from the shelf slope. Once the reflected wave is separated from the incident wave, both energy components remain constant. Since no wave breaking is induced during wave propagation over the shelf, there is no energy dissipation.

A similar behaviour of both energy components by using a RANS model was reported by Liu and Cheng (2001) for a single solitary wave propagating over a rectangular shelf placed on a horizontal bottom. The transfer of kinetic energy in potential energy (resulting in the increase of potential and the decrease of kinetic energy) is more noticeable over a flat bottom as in Liu and Cheng (2001). This is due to the effect of the sloping bottom in own simulations, which additionally causes wave shoaling and wave reflection.

b) Definition of transmitted, reflected and dissipated wave energy in case of reef structure present and nonbreaking waves

Transmitted wave energy per unit width in case of nonbreaking waves can be found by subtraction of the energy of the wave reflected by the reef from the incident wave energy (determined for no reef) as illustrated in Figure 5.5. Kinetic, potential and total energy are normalized by the value of the total wave energy computed at normalized time t(g/h)0.5=344.53, while time is normalized by t(g/h)0.5. Time t(g/h)0.5=0 corresponds to the entrance of a solitary wave into the numerical domain (i.e. the wave starts propagating over the shelf slope).

The energy pattern of the nonbreaking wave propagating over the shelf containing the reef follows the pattern of the wave travelling over the shelf only, until the wave approaches the reef. As the wave reflected by the reef leaves the numerical domain, the total energy decreases and remains constant once this process is accomplished. The reflected wave energy per unit width is calculated as the difference between the incident wave energy and the energy remaining after the exit of the reflected wave from the domain. There is no dissipated energy due to wave breaking. As a result of wave shoaling over the shelf slope, kinetic energy is transferred into potential energy. Once this process is accomplished, both energy components remain constant. As the wave approaches the reef, potential energy increases (while kinetic energy decreases) due to shoaling over the obstacle. Reflection of the wave from the reef causes a reduction of potential wave energy (i.e. its transfer into kinetic energy). Once the reflected wave is separated from the incident wave, both kinetic and potential energy remains constant. The total energy is reduced by the energy of the wave reflected from the reef, leaving the numerical domain through the left boundary.

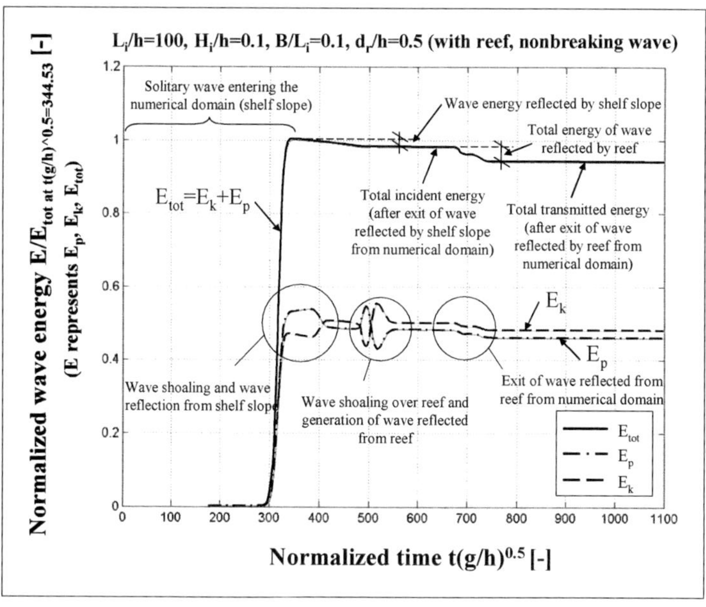

Figure 5.5: Determination of incident, transmitted, reflected and dissipated wave energy in case of reef present and nonbreaking wave

c) Definition of transmitted, reflected and dissipated wave energy in case of reef structure present and breaking waves

In case of breaking waves, the total energy per unit width is reduced twice: due to energy dissipation associated with wave breaking over the reef and due to the part of the energy reflected from the reef slopes that leaves the numerical domain (see Figure 5.6). Kinetic, potential and total energy are normalized by the value of the total wave energy computed at normalized time $t(g/h)^{0.5}=297.55$, while time is normalized by $t(g/h)^{0.5}$. Time $t(g/h)^{0.5}=0$ corresponds to the entrance of a solitary wave into the numerical domain (i.e. the wave starts propagating over the shelf slope).

The dissipated energy per unit width is determined as the difference between the incident wave energy (determined from simulations without the reef) and the energy remaining after completion of the breaking process. The energy per unit width reflected by the reef is computed as the difference between the energy remaining after wave breaking and the energy remaining after the exit of the wave reflected by the reef from the numerical domain. Consequently, the transmitted wave energy per unit width is determined as the difference between the energy remaining in the domain after the breaking process is completed and the energy remaining after the wave breaking and the exit of the wave reflected from the reef from the computational domain.

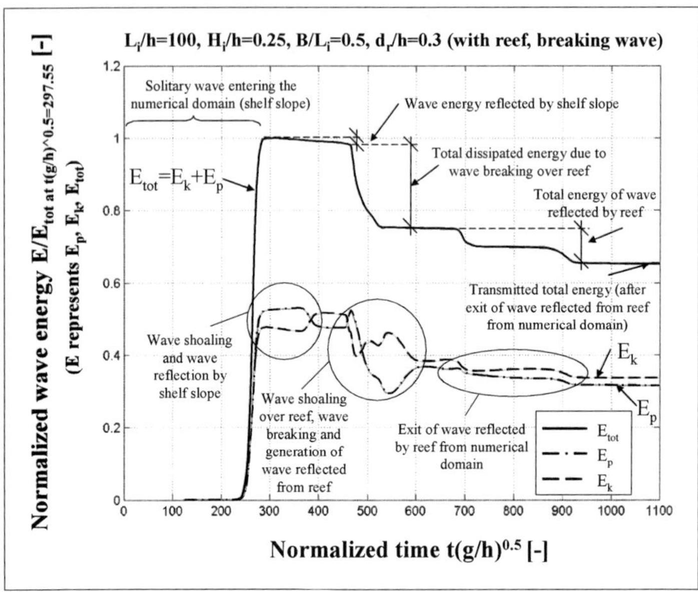

Figure 5.6: Determination of incident, transmitted, reflected and dissipated wave energy in case of reef present and breaking wave

The behaviour of both kinetic and potential energy is similar to that of the nonbreaking wave propagating over the shelf with the reef, until reaching the structure. Due to wave shoaling over the obstacle slopes, kinetic energy is transferred into potential energy. The inception of wave breaking over the reef causes a sudden reduction of both kinetic and potential energy. The energy pattern is further modified by the generation and the separation of the reflected wave from the incident wave. After termination of these processes, kinetic and potential energy remain constant first and are reduced later by the energy of the wave reflected from the reef that leaves the computational domain.

5.2.2 Analysis of numerical results

In this section, the damping performance of artificial reef under tsunami-like solitary wave conditions is investigated for varying structure geometry and incident wave parameters. Wave transmission, wave reflection and wave energy dissipation coefficients are computed to indicate the reef effectiveness in attenuation of the incident tsunami. Classification of the evolution modes of a solitary wave in simulations without and with a reef is required first to understand the processes controlling the damping functioning of the structure.

a) Classification of solitary wave evolution modes

The simulated evolution modes of a solitary wave transmitted over an artificial reef (see Tables F1-F4 in Appendix F) are similar to those observed in the laboratory experiments already described in Chapter 3 (see Figures 5.7-5.9):

- *Fission of a single incident nonbreaking wave* typical for simulations without reef on the horizontal shelf (i.e. for B/L_i=0.0, d_r/h=1.0) and all relative incident wave heights H_i/h (see Figure 5.7 and Tables F1-F4 in Appendix F). In the numerical simulations, the fission process starts over the horizontal shelf, in the region of the domain corresponding to the region in front of the submerged reef for cases with $B/L_i \neq 0$. Wave breaking is not generated during wave propagation both over the shelf slope and the horizontal shelf.

  A single incident solitary wave starts to disintegrate into solitons over the horizontal part of the shelf at a distance of approximately x/h=95 for H_i/h=0.10; x/h=57 for H_i/h=0.15; x/h=35 for H_i/h=0.20 and x/h=26 for H_i/h=0.25 measured from the beginning of the horizontal part of the shelf (see Figure 5.7a). The following number of the humps generated at the rear part of the single incident solitary wave (representing emerging solitons) is observed, including the leading wave (first soliton): N=4 for $H_i/h=0.10$, N=5 $H_i/h=0.15$, N=6 $H_i/h=0.20$ and N=7 $H_i/h=0.25$ (see Figure 5.7b). However, only one soliton for $H_i/h=0.10$, two solitons for $H_i/h=0.15$, three solitons for $H_i/h=020$ and 4 solitons for $H_i/h=0.25$ fully emerged (i.e. preserved the solitary wave shape) during the propagation distance over the shelf (see Figure 5.7c).

Solitary wave disintegration into solitons is well documented by the experimental and numerical studies available in the literature on solitary wave fission over submerged structures of infinite widths (e.g. Madsen and Mei, 1969; Seabra-Santos et al., 1987; Losada et al. 1989; Liu and Cheng, 2001). The results of own numerical simulations agree with those from the available studies – the wave scattering due to the water depth change when propagating from the deeper portion of water to the region of shallower water over the shelf is predicted according to the above mentioned investigations. The number of the solitons emerged in own simulations and the available studies cannot be however compared, since different ratio h/L_i was examined.

This type of evolution mode was not observed in own laboratory experiments described in Chapter 3, since for $B/L_i=0.0$ the waves propagated only over a horizontal bottom in the wave flume (i.e. there was no change in water depth required to generate the fission process).

- *Fission of transmitted nonbreaking wave* with the incident wave disintegrating into at least two solitons, trailed by oscillatory waves. No breaking of the leading soliton is induced, which is typical for small values of H_i/h and d_r/h (see Figure 5.8 and Tables F1-F4 in Appendix F).

No wave breaking occurs for the simulations with relative incident wave height $H_i/h=0.1$ for all considered relative reef widths B/L_i and relative submergence depths d_r/h. In case of $H_i/h=0.15$, no breaking is generated for all ratios d_r/h for $B/L_i=0.1$ and $d_r/h=0.4$, 0.3 for $B/L_i=0.2$-1.0. No breaking process is observed in case of $H_i/h=0.20$ for $d_r/h=0.4$, 0.5 for $B/L_i=0.1$ and $d_r/h=0.5$ for $B/L_i=0.2$-1.0.

The wave starts to disintegrate already in front of the reef (at the same distances as given for the simulations without the reef) due to the first water depth change when propagating from the deeper water region towards the shelf. However, only two humps (including the leading wave) are generated in front of the reef for $H_i/h=0.10$ and 0.15 and three humps for $H_i/h=0.20$ and 0.25 (see Figure 5.8a). The process of wave disintegration into solitons continues in the shallower water over the reef crest, with additional humps generated (see Figure 5.8b).

As the soliton wave train enters the deeper portion of water behind the reef, new humps appear, followed by oscillatory waves (see Figure 5.8c). During the propagation along the shelf only few first solitons fully emerged (i.e. preserved the solitary wave shape). The determination of the exact number of the humps generated over and behind the reef, including the effect of relative structure width B/L_i and relative submergence depth d_r/h on the number of the generated solitons, would require performance of very detailed simulations and analysis, which is outside the scope of this study.

As already mentioned in Chapter 2, there is no available systematic study on the fission of nonbreaking solitary wave due to a submerged structure of a finite width that could be used to compare with the results of own numerical simulations. This phenomenon was briefly reported by Lin (2004). Prediction of this evolution mode by own numerical simulations is similar to that of Lin (2004) (see Figure G1 in Appendix G).

This evolution mode of a solitary wave was also observed in own laboratory experiments as already discussed in Chapter 3.

- *Fission of transmitted wave followed by the breaking of leading soliton.* Fission of the incident wave, already initiated in front of the reef, develops further in the shallower water over the reef crest, where the leading soliton becomes unstable and starts to break (see Figure 5.9 and Tables F1-F4 in Appendix F).

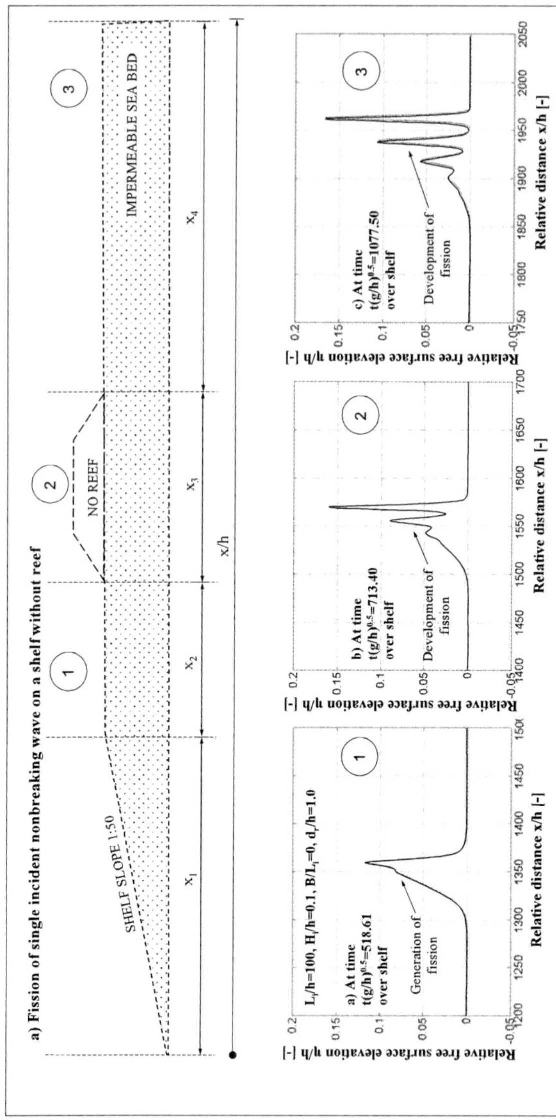

Figure 5.7: Modes of solitary wave propagation in simulations - fission of single incident nonbreaking wave over shelf without artificial reef

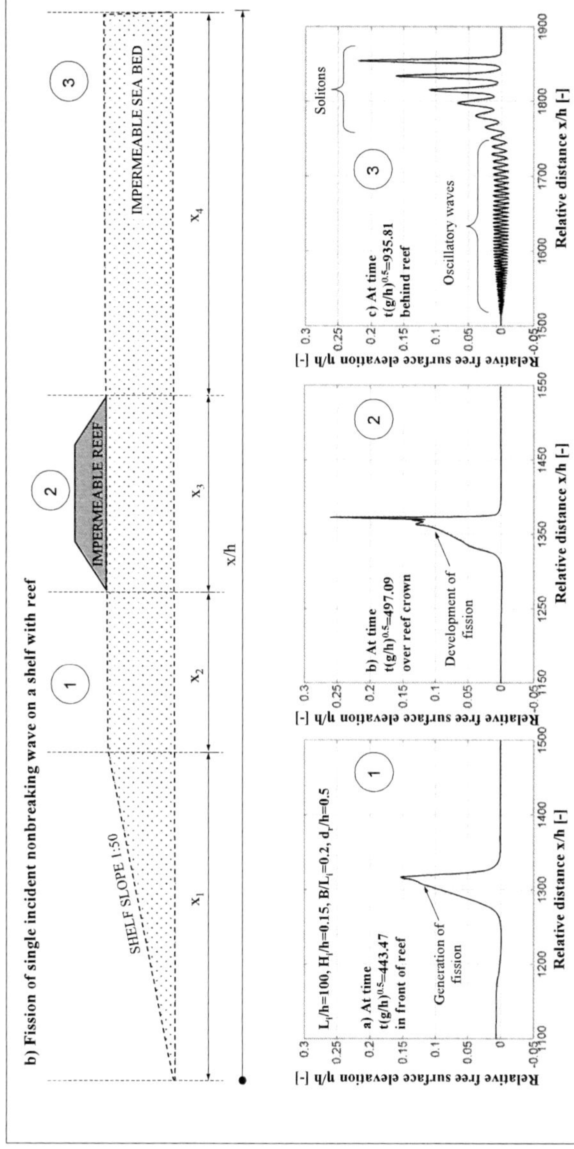

Figure 5.8: Modes of solitary wave propagation in simulations - fission of transmitted nonbreaking wave over shelf with artificial reef

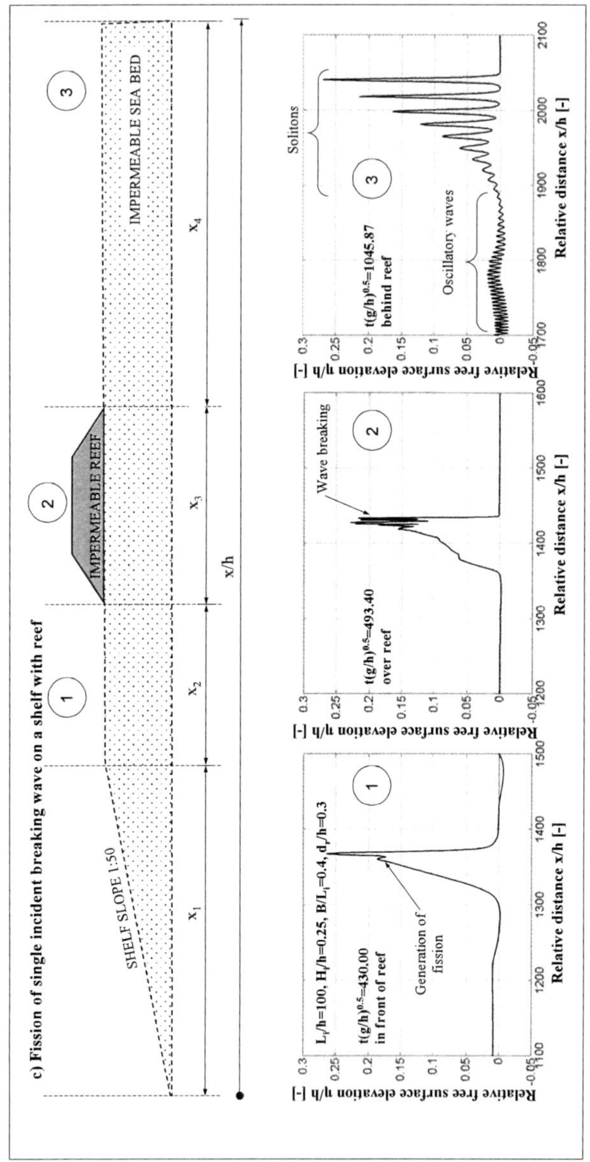

Figure 5.9: Modes of solitary wave propagation in simulations - fission of transmitted wave followed by wave breaking over shelf with artificial reef

For the relative incident wave height H_i/h=0.15, incipient wave breaking of the leading soliton occurs at a distance of about 21 m behind the seaward reef toe only for the smallest submergence depth d_r/h=0.3 and relative reef widths B/L_i=0.2-1.0. For the relative wave height H_i/h=0.20, breaking occurs for the two smallest considered relative submergence depths d_r/h=0.3 and 0.4 for the relative reef widths B/L_i=0.2-1.0, while for B/L_i=0.1 it occurs only for d_r/h=0.

The leading soliton starts to break over the reef crest at a distance of about 20 m from the seaward reef toe for d_r/h=0.4 and about 13 m from the seaward reef toe for d_r/h=0.3. For the relative wave height H_i/h=0.25, the leading soliton is forced to break over the reef crown for all the considered reef geometries (apart from the case of B/L_i=0.1, d_r/h=0.5) at a distance of about: 23 m for d_r/h=0.5, 12 m for d_r/h=0.4 and 10 m for d_r/h=0.3 from the seaward reef toe. No breaking of the incident wave is observed in front of the reef, so that the other evolution mode observed in the laboratory experiments (breaking of the transmitted wave followed by wave fission of the broken wave) cannot be observed.

It can be concluded that the breaking conditions over the reef are governed primarily by relative incident wave height H_i/h and relative reef submergence depth d_r/h. The increase of wave nonlinearity from H_i/h=0.1 to 0.25 causes more favourable conditions for the generation of wave breaking.

Extension of the relative width of the reef crest leads to shifting of the breaking triggering point towards the seaward reef toe, what can be also observed for a given B/L_i ratio and decreasing d_r/h ratio. Wave breaking event induced over the reef crest is terminated once the wave enters the deeper portion of water behind the structure.

Similar to the above described evolution mode (fission of transmitted nonbreaking wave), incident wave starts to disintegrate into solitons already in front of the reef (at the same distances as given for the simulations without the reef). This is due to the first water depth change when propagating from the deeper water region towards the shelf.

However, only two humps (including the leading wave) are generated in front of the reef for H_i/h=0.10 and 0.15 and three humps for H_i/h=0.20 and 0.25 (see Figure 5.9a). The process of wave disintegration into solitons continues in the shallower water over the reef crest, with additional humps generated (see Figure 5.9b) till generation of breaking process. Once the breaking event is completed, further wave disintegration (together with the generation of oscillatory waves) occurs as the wave train enters the deeper water behind the reef.

According to the literature review presented in Chapter 2, there is no available systematic study on the fission of breaking solitary wave due to a submerged structure of a finite width that could be used to compare with the results of own numerical simulations. A brief remark on such wave behaviour was provided by Lin (2004): "For the breaking waves (...) the leading wave is followed by a trailing edge whose amplitude reduces at first but increases later". The wave pattern predicted numerically by own simulations is similar to that by Lin (2004) (see Figure G2 in Appendix G).

Similarly to the performed laboratory experiments, the solitary wave fission is not suppressed by wave breaking, thus supporting the conclusions of Liu and Cheng (2001). The comparison of the breaking conditions in the performed numerical and the laboratory experiments reveals that the leading solitons started breaking always over the reef crown in the simulations, while in the laboratory tests close to the landward reef corner or the landward reef slope or behind the reef, depending on incident wave conditions and structure geometry.

b) Determination of wave transmission, wave reflection and wave energy dissipation coefficients

The hydraulic performance of impermeable artificial reef of different configurations is investigated for varying incident conditions of a solitary wave-like tsunami. The determination of the rate of wave transmission, wave reflection and wave energy dissipation in terms of the coefficients of wave transmission K_t, wave reflection K_r and wave energy dissipation K_d is required to indicate explicitly the effectiveness of the structure in attenuation of tsunami waves. Moreover, a feasibility study of the artificial reef is performed to specify the structure dimensions required for effective tsunami damping.

As already mentioned, the coefficients K_t, K_r, K_d are determined in terms of wave energy according to Eqs. 2.14 and 2.15 due to the generation of the fission process, with the transmitted, reflected and dissipated energy components as defined in Section 5.2.1 (see Tables F1-F4 in Appendix F).

The calculated coefficients K_t, K_r, K_d are plotted versus relative structure width B/L_i in Figures 5.10-5.12, respectively. Curves are drawn for different relative wave heights H_i/h and different relative submergence depths d_r/h, and a distinction is made between the nonbreaking and breaking wave conditions over the reef. The main conclusions of the analysis of wave transmission, wave reflection, and wave energy dissipation for simulations with and without the reef placed on the shelf are summarized below.

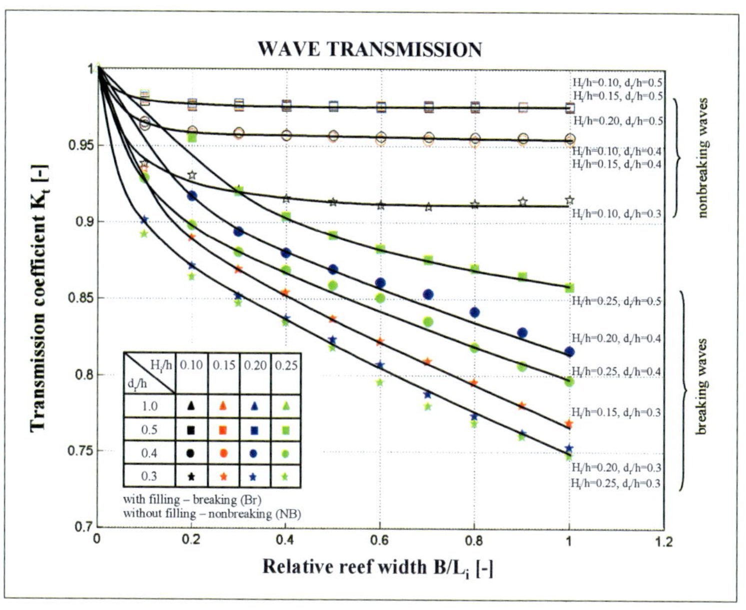

Figure 5.10: Wave transmission versus relative structure width B/L_i for different relative incident wave heights H_i/h and different relative submergence depths d_r/h

Nonbreaking waves, without artificial reef on the shelf

- Full wave transmission (K_t=1.0) occurs for the simulations without the reef on the horizontal shelf (i.e. for B/L_i=0.0 and d_r/h=1.0) as shown in Figure 5.10. As a consequence, there is no wave reflection (K_r=0.0) that would normally take place at the seaward and landward slopes of the reef as indicated in Figure 5.11. Additionally, there is no energy dissipation (K_d=0.0) for nonbreaking wave conditions over the horizontal shelf as shown in Figure 5.12.

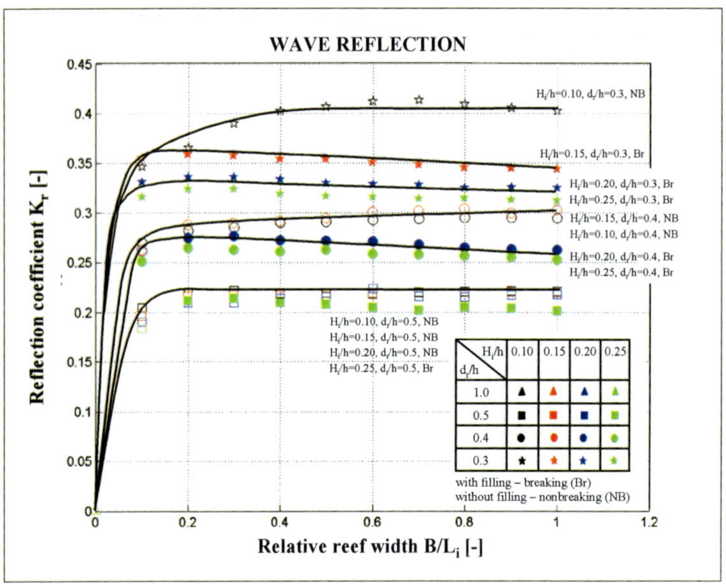

Figure 5.11: Wave reflection versus relative structure width B/L_i for different relative incident wave heights H_i/h and different relative submergence depths d_r/h

Nonbreaking waves, with artificial reef on the shelf

- No decrease of wave transmission coefficient can be observed with the elongation of the reef from B/L_i=0.1 to B/L_i=1.0. This would indicate that the effect of the structure presence only on wave attenuation is negligible.

For a given relative submergence d_r/h, transmission coefficient is approximately of the same magnitude irrespective of the relative incident wave height and the relative reef width: K_t~0.98 for d_r/h=0.5, K_t~0.96 for d_r/h=0.4 and K_t~0.92 for d_r/h=0.3. For a given relative incident wave height H_i/h, wave transmission decreases with the decrease of the relative submergence depth (see Figure 5.10).

- The only source of wave energy reduction is due to wave reflection from the reef slopes. Similarly to the transmission coefficient, reflection coefficient is approximately constant for almost all considered relative structure widths (see Figure 5.11). Moreover, it is also constant for a given submergence depth d_r/h regardless of the incident wave conditions: K_r~0.23 for d_r/h=0.5, K_r~0.29 for d_r/h=0.4 and K_r~0.41 for d_r/h=0.3. For a given incident wave height H_i/h, the rate of wave reflection increases with the decrease of the submergence depth.
- Due to nonbreaking wave conditions over the reef, there is no energy dissipation (K_d=0.0) as shown in Figure 5.12.

Breaking waves, with artificial reef on the shelf

- Wave breaking significantly contributes to the reduction of wave transmission. Since the broken wave propagates as a progressive turbulent bore, the energy losses become significant with increasing relative structure width B/L_i (the breaking event stops immediately once wave enters the deeper water behind the reef).
- For a given B/L_i ratio and incident wave height H_i/h, the transmission coefficient decreases with the reduction of relative water depth over the reef d_r/h, since more favourable conditions for generation of breaking event are created (see Figure 5.10). For a given relative structure width B/L_i and relative submergence depth d_r/h, wave transmission is reduced as wave nonlinearity H_i/h increases. The minimum of wave transmission (K_t~0.75) is achieved for the widest reef (B/L_i=1.0), the smallest submergence depth (d_r/h=0.3) and the higher incident wave height (H_i/h=0.25).

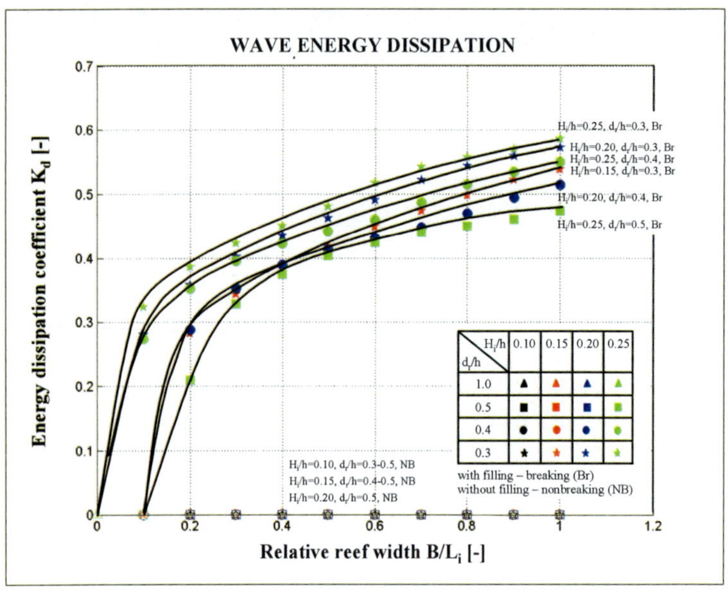

Figure 5.12: Wave energy dissipation versus relative structure width B/L_i for different relative incident wave heights H_i/h and different relative submergence depths d_r/h

- For a given d_r/h ratio, the reflection coefficient increases with the elongation of the structure width up to B/L_i=0.2 and remains approximately constant for B/L_i>0.2 (see Figure 5.11). For a given B/L_i ratio and incident wave height H_i/h, the reflection coefficient increases with the reduction of water depth over the reef d_r/h, that corresponds to the increase of relative structure height h_r/h. For a given relative structure width B/L_i and relative submergence depth d_r/h, wave reflection decreases as wave nonlinearity H_i/h increases. The maximum wave reflection (K_r~0.35) is achieved for the widest structure (B/L_i=1.0), smallest submergence depth (d_r/h=0.3) and relatively small incident wave height (H_i/h=0.15).
- For a given d_r/h ratio, the energy dissipation coefficient increases with the elongation of the structure width (see Figure 5.12). For a given B/L_i ratio and incident wave height H_i/h, the energy dissipation coefficient increases with the reduction of water depth over the reef d_r/h, since more favourable

conditions for inception of wave breaking are generated. For a given relative structure width B/L_i and relative submergence depth d_r/h, the rate of wave energy dissipation increases with the increase of wave nonlinearity H_i/h. The maximum wave energy dissipation (K_d~0.59) is achieved for the widest structure (B/L_i=1.0), smallest submergence depth (d_r/h=0.3) and highest incident wave height (H_i/h=0.25).

The results of the numerical analysis on the hydraulic performance of artificial reef for tsunami conditions do not generally indicate existence of optimal reef width (required only to force wave to break) as reported in study of Dattatri et al. (1978) on damping storm waves due to submerged structures (here B/L_i=0.2-0.3 was suggested) or by Yoshioka et al. (1993) with the optimal relative structure width B/L_i=1.5. Such trend can be observed only for nonbreaking waves over the reef, for which the transmission coefficient becomes approximately constant for B/L_i>0.2. In case of waves broken over the structure, the rate of wave transmission decreases significantly and continuously with the widening of relative structure width B/L_i, since dissipation of wave energy due to breaking event accompanies wave propagation over structure crest and is finished once the wave enters the deeper portion of water behind the structure.

The effect of presence of an artificial reef on the attenuation of nonbreaking waves is rather negligible, since the only source of wave energy reduction is due to the reflection of the incident wave from the structure slopes (as already mentioned, the minimum wave transmission yields K_t~0.92). The improvement of the damping performance of the reef can be achieved in this case solely by decreasing the relative submergence depth (i.e. by increasing the relative structure height).

A significant reduction of incident wave energy can be observed solely for waves breaking over the reef crest, since the turbulent processes associated with the breaking event, contribute additionally to the wave energy dissipation. In this case, the damping performance of the reef is determined by both relative submergence depth d_r/h and relative structure width B/L_i and can be improved by decreasing the ratio d_r/h and increasing the ratio B/L_i. However, even for the widest structure and the smallest submergence depth (i.e. for B/L_i=1.0, d_r/h=0.3), the transmission coefficient is reduced only by about 25% (K_t~0.75). Considering the construction costs of such an artificial reef (of a crest width that is approximately of the same length as the length of an incident wave) in respect to the predicted wave energy reduction, application of the reef itself as a protection against tsunami is indeed questionable. However, the contribution of the artificial reef to the overall damping performance of the multi-defence line strategy against tsunami (see Figure 1.1) should be further examined, since the solitary wave train generated due to wave fission can be more easily attenuated by the next defence lines in comparison to a single incident tsunami wave.

5.3 Summary and implications of the numerical results

The numerical investigations of the hydraulic performance on an artificial reef include the analysis of the local processes (wave breaking and wave fission) and global processes (wave transmission, wave reflection and wave energy dissipation). Due to the research gap on the nonlinear behaviour of a tsunami-like solitary wave over a submerged structure of a finite width, understanding of the local processes, determining the damping functioning of the structure, is required. Modes of solitary wave evolution are classified and compared both to those observed in own laboratory experiments and reported in the previous studies. The analysis of the global processes indicates the effectiveness of the reef to attenuate tsunami-like solitary wave and allows to determine reef dimensions corresponding to the most effective wave damping. The coefficients of wave transmission, wave reflection and wave energy dissipation are determined for a solitary-like tsunami wave of different incident characteristics (relative wavelength L_i/h=100, relative wave heights H_i/h=0.10-0.25) propagating over an impermeable submerged structure of varying geometry (relative width B/L_i=0.0-1.0, relative submergence depth d_r/h=0.3, 0.4, 0.5, 1.0, seaward and landward slopes 1:10), for an arbitrary water depth *h*. The key results of the numerical experiments can be summarized as follows:

- A brief description of wave breaking conditions over the reef is provided and compared to the results from own laboratory tests and the available studies. Breaking event is always induced over the reef crest (with the inception point shifted towards the seaward upper reef corner with increasing wave nonlinearity H_i/h) and results from the amplification of the incident wave height due to shoaling over the seaward structure slope. The process generally occurs for higher relative incident wave heights (H_i/h=0.20, 0.25) and smaller relative submergence depths (d_r/h=0.3, 0.4). No breaking is generated during wave propagation over the shelf without the reef.

- A brief characteristic of the fission process is provided and compared with the results from own laboratory tests and the available studies. Wave disintegration into a solitary wave train (i.e. solitons of decreasing heights) followed by oscillatory waves is generated during wave propagation over the shelf without the reef due to the water depth reduction over the shelf. In case of wave propagation over the shelf with the reef, additional humps representing solitons are generated over the structure crest as well as in the deeper water behind the reef. During the propagation distance however, only the first few waves are fully developed (i.e. preserve solitary wave shape).
- The effect of wave breaking (and the associated energy dissipation) on wave transmission is found to be significant and increases with increasing relative structure width B/L_i. For a given relative incident wave height H_i/h, the minimum wave transmission is always attributed to the largest relative structure width (B/L_i=1.0) and the smallest relative submergence depth (d_r/h=0.3), and it decreases with the increasing incident wave nonlinearity due to the induced energy dissipation accompanying wave breaking: K_t~0.92 for H_i/h=0.10 (nonbreaking wave), K_t~0.77 for H_i/h=0.15 (breaking wave), K_t~0.75 for H_i/h=0.20 (breaking wave), K_t~0.75 for H_i/h=0.25 (breaking wave).
- For nonbreaking wave conditions, the transmission performance is low (K_t=0.92-0.98) and widening of the structure crest does not result in a further reduction of wave transmission. An optimal relative structure width is clearly noticeable in this case and yields B/L_i=0.2-0.4 – beyond these values the transmission coefficient remains unchanged: K_t~0.98 for H_i/h=0.10-0.20, d_r/h=0.5; K_t~0.96 for H_i/h=0.10-0.15, d_r/h=0.4 and K_t~0.92 for H_i/h=0.10, d_r/h=0.3.
- For breaking wave conditions there is no clear optimal relative structure width, since the broken wave resembles a progressive bore, dissipating wave energy continuously until entering the deep water behind the reef.

Consequently, the numerical results do not indicate the existence of an optimal relative structure width of value B/L_i=0.2-0.3 as postulated by Dattatri et al. (1978) or B/L_i=1.5 by Yoshioka et al. (1993) for damping of storm waves by submerged structures.

- In case of simulations without the reef on the shelf (nonbreaking wave conditions), full transmission of the incident waves is observed (K_t=1.0).
- The rate of wave reflection tends to be constant with increasing relative structure width B/L_i for both nonbreaking and breaking waves and ranges from K_r~0.2 to K_r~0.4. The reflection coefficients increases as relative incident wave height H_i/h becomes smaller and relative submergence depth d_r/h becomes smaller. No reflection (K_r=0.0) is observed for waves propagating over the shelf without the reef.
- For nonbreaking waves there is no dissipation of wave energy due to the turbulent processes associated with the breaking event (K_d=0.0). For breaking waves, the coefficient of wave energy dissipation increases with the increase of relative structure width B/L_i, increase of relative incident wave height H_i/h and decrease of relative submergence depth d_r/h. Since wave energy is dissipated continuously over the reef crown, the largest coefficient of energy dissipation is observed for the widest structures (B/L_i=1.0) K_d~0.48 to K_d~0.59.
- The effectiveness of artificial reef to attenuate a solitary-like tsunami wave is indicated only for waves broken over the structure crest. However, even in this case the maximum reduction of the incident wave is relatively low (only about 25%) as compared to the structure dimensions required to achieve this rate of wave attenuation (B/L_i=1.0, d_r/h=0.3). Considering the construction costs of such a structure (of width equal approximately the length of an incident tsunami wave), application of artificial reef to reduce tsunami impact on a shore is rather impractical. However, a study of the effect of the damping functioning of the reef on the overall performance of the multi-defence line protection system against a tsunami would be required.

6. Summary, conclusions and outlook

The main results and conclusions drawn from this study are summarized in this chapter. Moreover, limitations of the results are provided and the priority tasks for the future research are identified.

6.1 Summary of key results and conclusions

Considering the urgent need to establish a reliable defence system against extreme tsunami that emerged after the 2004 event, the application of an artificial reef as a first defence line to attenuate incident tsunami energy within the multi-defence line strategy is investigated in this study. Since the functioning of the entire defence system will depend on the performance of this first line, the main objective of the research is therefore to provide the damping performance of the reef structure in terms of wave transmission, wave reflection and wave energy dissipation for different incident wave conditions (relative water depth h/L_i and relative wave height H_i/h) and reef parameters (relative structure width B/L_i and relative submergence depth d_r/h). As a result, the required structure dimensions to achieve the most effective wave attenuation can be determined.

Beside a systematic analysis of the present state of knowledge and modelling, the study consists of two important phases: (i) laboratory experiments in the 2m wide wave flume of LWI on the nonlinear transformation of a solitary wave over a submerged structure of a finite width, (ii) numerical investigations using a modified model COULWAVE (developed by Lynett, 2002) on the hydraulic performance of artificial reef subjected to a tsunami-like solitary wave.

6.1.1 Laboratory experiments on nonlinear transformation of solitary wave over impermeable submerged structure of finite width

Extensive and systematic laboratory tests are performed for the first time to provide a description of solitary wave evolution modes when interacting with a submerged impermeable structure of a finite width, including both breaking and nonbreaking wave conditions for different water depths (h=0.5, 0.6, 0.7 m), incident wave heights ($H_{i,nom}$=0.06-0.22 m), structure heights (h_r=0.0, 0.3, 0.4, 0.5 m) and structure widths (B=0.0, 1.0, 2.0 m). The effect of structure shape (trapezoidal and rectangular), relative incident wave height H_i/h, relative structure width B/L_i and relative submergence depth d_r/h on incipient wave breaking and wave fission of a solitary wave is examined. A wave breaking criterion was developed and the breaker types are classified on the basis of the studies available in the literature. The number of the solitons which emerge in the process of wave disintegration for nonbreaking waves is estimated in comparison to the studies available in the literature on solitary wave fission over submerged structures of infinite widths.

The key results from these laboratory investigations can be summarized as follows:

Analysis of solitary wave evolution modes

The following solitary wave evolution modes over the impermeable submerged structure of a finite width are observed and confirmed by the available studies related to nonlinear transformation of solitary wave over structures of infinite widths: (i) no fission of a single incident nonbreaking wave, (ii) weak distortion of a single transmitted nonbreaking wave, (iii) fission of a single transmitted nonbreaking wave, (iv) fission of breaking leading wave (with two cases distinguished: fission of a single transmitted wave followed by the breaking of the leading soliton with further development of the fission process; breaking of a single transmitted wave followed by wave fission of the broken wave with further evolution of the fission process).

Analysis of wave breaking

- The observed wave breaker types (spilling, transition and plunging breakers) were confirmed by the wave breaker classification reported in the previous studies on solitary wave breaking over submerged impermeable structures.
- Incipient breaking of a solitary wave is dependant on relative structure width B/L_i and relative wave height H_i/d_r, and independent of the structure shape. Accordingly a breaking criterion is developed, which is valid for B/L_i=0.2-0.49 and H_i/d_r=0.2-2.37:

$$H_i/d_r = 0.573\ (B/L_i)^{-0.6}$$

- Inception of wave breaking over the structure does not suppress the generation of soliton fission.

Analysis of wave fission

- Disintegration of a single incident solitary wave in a train of waves (at least two) of decaying heights (solitons), followed by oscillatory waves, is observed as a result of wave interaction with the submerged structure.
- Similarly to the previous studies on wave fission over submerged structures of infinite widths, the number of the emerged solitons for nonbreaking waves is not dependant on relative incident wave height H_i/h for a given relative submergence depth d_r/h.
- The parameters influencing the number of the generated solitons are relative submergence depth d_r/h and relative structure width B/L_i, as shown in Table 6.1. The number of solitons increases with decreasing relative submergence depth d_r/h and increasing relative structure width B/L_i. More systematic and comprehensive laboratory tests will be needed to come up with a generic relationship between the number of solitons N and the reef parameters d_r/h and B/L_i. The observed solitons number (under nonbreaking wave conditions for water depth h=0.6m) for the smaller reef width.
- The solitons number observed in the laboratory experiments is close to that reported in the studies on solitary wave fission over submerged structures of infinite widths, and it clearly indicates that relative structure width B/L_i is also important.

Table 6.1: Solitons number observed in own laboratory experiments

B [m]	$B/L_{i,nom}$ [-]	d_r/h [-]		
		0.17	0.33	0.50
1.0	0.12-0.24	4	3	2
2.0	0.25-0.48	5	4	3

6.1.2 Numerical experiments on hydraulic performance of artificial reef under tsunami impact

Systematic numerical investigations using a modified numerical model for water wave propagation COULWAVE (developed by Lynett, 2002) is performed to provide the damping performance of an impermeable artificial reef under tsunami-like solitary wave conditions. Wave transmission, wave reflection and wave energy dissipation coefficients are determined in terms of wave energy – the most suitable method in case of generation of wave fission accompanying wave transmission over a submerged structure. For this purpose, formulae for the calculation of kinetic and potential energy of a solitary wave are derived and implemented in the source code.

Unlike most of the available investigations, in which a short-wavelength/period tsunami wave is generated over a horizontal bottom, more realistic tsunami conditions are modelled (with h/L_i=0.01 typical for long waves). This is achieved by generating a solitary wave at a location with deeper water from which the wave propagates towards the shelf containing the artificial reef.

It should be underlined, that such analysis of the damping performance of an artificial reef cannot be conducted under laboratory conditions due to the limitations of the horizontal-vertical scale, which does not allow at the present stage to generate realistic (i.e. long-period) tsunami waves.

The main aim of the numerical study is to investigate the applicability and feasibility of an artificial reef for coastal protection against tsunami by providing the reef dimensions required to achieve the maximum wave damping for given incident wave conditions (H_i/h=0.1-0.25; L_i/h=100; h=1.0 m) and reef parameters (d_r/h=0.3-0.5,1.0; B/L_i=0-1.0; sea- and landward slopes 1:10).

The key results from this investigation can be summarized as follows:

- A significant reduction of the incident wave energy can be achieved only through forcing the incident wave to break over the structure. In this case, no optimal structure width can be identified as reported in the studies on storm wave attenuation, since the wave over the structure resembled a turbulent bore dissipating the energy until entering the deep water behind the reef (i.e. the wider the structure, the smaller the transmission coefficient). Apart from relative structure width B/L_i, the other parameters controlling wave transmission are relative incident wave height H_i/h and relative submergence depth d_r/h. The transmission coefficient is reduced as the ratio H_i/h increases and as the ratio d_r/h decreases. The minimum wave transmission is achieved for the widest structure (B/L_i=1.0) and the smallest submergence depth (d_r/h=0.3), yielding: K_t=0.769 for H_i/h=0.15, K_t=0.753 for H_i/h=0.20 and K_t=0.747 for H_i/h=0.25.
- For nonbreaking conditions (H_i/h=0.10), the transmission coefficient decreases with the relative structure width until value B/L_i=0.2 for d_r/h=0.4, 0.5 and B/L_i=0.4 for d_r/h=0.3 and then becomes constant ($K_t\approx$0.975 for d_r/h=0.5, $K_t\approx$0.956 for d_r/h=0.4 and $K_t\approx$0.911 for d_r/h=0.3). Unlike the relative submergence depth, the effect of the relative wave height H_i/h on wave transmission is negligible unlike the relative submergence depth. The transmission coefficient decreases with decreasing water depth over the structure (the minimum transmission is achieved for the smallest submergence depth d_r/h=0.3).

6.2 Applicability of the results, limitations and outlook

The results obtained from this study can be applied to both engineering practice and further investigations on this topic, as specified below.

a) Applicability and limitations of the results of the laboratory experiments

The wave breaking criterion, developed on the basis of the laboratory results in Chapter 3, can be used to predict breaking conditions for a solitary wave over a submerged impermeable structure of a finite width and arbitrary shape. The formula (see Fig. 3.15) is applicable however for the following range of relative structure width B/L_i and relative wave height H_i/d_r: B/L_i=0.2-0.49; H_i/d_r=0.2-2.37.

Experiments on breaking conditions over a very narrow structure (i.e. $B/L_i \rightarrow 0$) and over a structure of an infinite width (i.e. $B/L_i \rightarrow \infty$) are not performed and would be needed to provide breaking criterion valid for an arbitrary ratio B/L_i.

The knowledge on wave fission provided by of the laboratory results described in Chapter 3, allows to estimate the number of the emerged solitons resulting from the propagation of a solitary wave over an impermeable submerged structure of a finite width. The soliton numbers are however determined solely for nonbreaking waves. The distinction of the solitons for broken waves is very difficult due to a complex wave profile after the inception of the breaking process and would require a performance of additional investigations. The other problem faced during the experiments is the distinction between the emerged solitons and the oscillatory waves following the solitons. The method adopted for this purpose is based on very dense measurements of the wave profile by wave gauges, whose location is adjusted to each new test. This allowed to follow generation of each hump at a rear part of a single wave, however confirmation of the provided soliton numbers by numerical investigations would be necessary.

The results of the laboratory experiments can be used to support further studies on the nonlinear transformation of a solitary wave at a submerged structure of a finite width, as suggested in Section 6.2c.

b) Applicability and limitations of the results of the numerical experiments on a hydraulic performance of an artificial reef under tsunami conditions

The provided analysis of wave transmission, wave reflection and wave energy dissipation can be used to predict the hydraulic performance of a submerged impermeable reef to be applied as a coastal protection against tsunami in an arbitrary water depth. Moreover, the results can support the optimization process of the reef geometry to achieve the most effective damping of incident tsunami energy.

In the simulations, dissipation of wave energy due to breaking process is considered only. Contribution of vortex shedding (around 10% according to the available literature) and bottom friction is not taken into account.

Under the following incident wave conditions: relative incident wavelength $L_i/h=100$, relative incident wave height $H_i/h=0.1\text{-}0.25$, impact of weakly nonlinear to nonlinear waves is examined, without the consideration of extreme tsunami (i.e. wave of larger ratios L_i/h and H_i/h). To provide a comprehensive attenuation performance of an artificial reef, simulations of very small relative submergence depth (i.e. $d_r/h \rightarrow 0$) should be included, since $d_r/h=0.3-0.5$, 1.0 are considered in the study.

Since the damping performance of the reef is determined for 1HD wave propagation, the effect of wave diffraction and current generation on wave transmission is not simulated.

The results of the numerical investigations can be used to support further studies on the hydraulic performance of an artificial reef under tsunami conditions as well as on the development of a multi-defence line strategy against tsunami, as suggested in Section 6.2c.

c) Outlook

This study has contributed to an improved understanding and prediction of the processes generated in the vicinity of a submerged impermeable reef of a finite width subject to a solitary-like tsunami impact. Both global processes (wave transmission, wave reflection and wave energy dissipation) and local processes (wave breaking and wave fission) are investigated. The research should be however considered as a first stage of a more systematic study on the applicability of an artificial reef as a defence against tsunami. In order to improve the determination of the effectiveness of the structure in attenuation of a tsunami, the following issues should be included in future research:

- Performance of additional experiments on solitary wave breaking over both very thin submerged structures (i.e. for $B/L_i \rightarrow 0$) and very broad barriers (i.e. for $B/L_i \rightarrow \infty$) in order to extend the application range of the determined breaking criterion.
- Improvement of the description of the fission mechanism in case of breaking wave conditions, particularly the determination of the soliton number. This can be achieved by a further development of the introduced visual method for the determination of the number of the emerged solitons. The method should allow to distinguish explicitly among the generated solitons and the oscillatory waves following the solitons, under both nonbreaking and

breaking wave conditions. Moreover, the description of the fission mechanism can be improved by performing numerical investigations by means of detailed flow models (such as COBRAS).

- Performance of laboratory experiments to determine the number of the emerged solitons in case of breaking waves over the submerged structure.
- Performance of numerical simulations for other reef geometries ($0<d_r/h<0.3$) and incident wave conditions ($H_i/h>0.25$) by employing a hybrid numerical model (i.e. coupled Boussinesq-RANS model) for water wave propagation to provide a total characteristic of the damping functioning of an artificial reef.
- Inclusion of energy losses both due to vortex shedding, flow through the structure (structure permeability) and bottom friction to improve the provided damping characteristic of an artificial reef by using a detailed flow model requiring low computational effort (i.e. hybrid models such as coupled Boussinesq-RANS model).
- Performance of 2HD simulations to examine the effects of wave diffraction and wave refraction on the hydraulic functioning of an artificial reef.
- Improvement of the characteristic of the damping performance of an artificial reef by calculating wave forces/pressure exerted by the transmitted wave on a structure placed behind the reef, representing e.g. house buildings onshore. This method would allow to indicate directly the effectiveness of an artificial reef in damping of an incident tsunami wave.
- Conducting a study on both structure stability under tsunami wave attack and scouring around the reef.

References

Agnon, Y.; Madsen, P.A.; Schäffer, H.A. (1999): A new approach to high-order Boussinesq models. *Journal of Fluid Mechanics*, No. 399, pp. 319-333.

Ahrens, J.P. (1987): *Characteristics of reef breakwaters*. Technical Report CERC-87-17, Coastal Engineering Research Center, U.S. Army Corps of Engineers Waterways Experiment Station, Vicksburg, USA, 62 pp.

Battjes, J.A. (1974): Surf similarity. *Proceedings of 14th International Conference on Coastal Engineering*, Kopenhagen, Denmark, pp. 466-480.

Beach Erosion Board (1940): *A model study of the effect of submerged breakwaters on wave action*. Technical Memo 1, Corps of Engineers, Washington DC, USA, 38 pp.

Beji, S.; Battjes, J.A. (1993): Experimental investigations on wave propagation over a bar. *Coastal Engineering*, No.19, pp. 151-162.

Benjamin, T.B.; Bona, J.L.; Mahony, J.J. (1972): Model equations for long waves in non-linear dispersive systems. *Phil. Trans. Roy. Soc. London A*, No. 272, pp. 249-274.

Bleck, M. (2003): *Hydraulische Wirksamkeit künstlicher Riffe am Beispiel einer rechteckigen Struktur*. Ph.D. Thesis, Leichtweiß-Institute for Hydraulic Engineering and Water Resources, TU Braunschweig, Germany (in German), pp. 144, www.biblio.tu-bs.de.

Boussinesq, J. (1872): Theorie des onde et des resous qui se propagent le long d`un canal rectangulaire horizontal, en communiquant au liquide contenu dans ce canal des vitesses sensiblement pareilles de la surface au fond. *Journal de Math. Pures et Appl.*, Deuxieme Serie, Vol. 17, pp. 55-108 (in French).

Brocchini, M.; Drago, M.; Iovenitti, L. (1992): The modelling of short waves in shallow waters. Comparison of numerical models based on Boussinesq and Serre equations. *Proceedings of 23th International Conference on Coastal*

Engineering, Venice, Italy, pp. 76-88.

Brocchini, M.; Dodd, N. (2008): Nonlinear shallow water equation modelling for coastal engineering. *Journal of Waterway, Port, Coastal, and Ocean Engineering*, Vol. 134, No. 2, pp. 104-120.

Bryant, E. (1994): *Tsunami. The underrated hazard*. Publisher Cambridge University Press, 320 pp.

Camfield, F.E. (1980): *Tsunami engineering*. Special Report No. 6., U.S. Army, Corps of Engineers, Coastal Engineering Research Center, Fort Belvoir, USA, 222 pp.

Chang, K.-A.; Hsu, T.-J.; Liu, P.L.-F. (2001): Vortex generation and evolution in water waves propagating over a submerged rectangular obstacle. Part I. Solitary waves. *Coastal Engineering*, No. 44, pp. 13-36.

Cooker, M.J.; Peregrine, D.H.; Vidal, C.; Dold, J.W. (1990): The interaction between a solitary wave and a submerged semicircular cylinder. *Journal of Fluid Mechanics*, No. 215, pp. 1-22.

d`Angremond, K.; Van der Meer, J.W.; De Jong, R.J. (1996): Wave transmission at low-crested structures. *Proceedings of the 25th International Conference on Coastal Engineering*, Orlando, Florida, pp. 2418-2427.

Dattatri, J.; Raman, H.; Jothi Shankar, N. (1978): Performance characteristics of submerged breakwaters. *Proceedings of the 16th International Conference on Coastal Engineering*, Hamburg, Germany, pp. 2153-2171.

Daemrich, K.F.; Mai, S.; Ohle, N. (2001): Wave transmission at submerged breakwaters. *Proceedings of the 4th International Symposium on Ocean Wave Measurement and Analysis*, San Francisco, USA, pp. 1725-1734.

Dean, W.R. (1945): On the reflection of surface waves by a submerged plane barrier. *Proceedings of the Cambridge Philosophic Society*, pp. 231-238.

Dean, R.G.; Dalrymple, R.A. (1991): *Water wave mechanics for engineers and scientists*. In Advanced Series on Ocean Engineering, Vol. 2, World Scientific, 353 pp.

Dick, T.; Brebner, A. (1968): Solid and permeable submerged breakwaters. *Proceedings of the 11th International Conference on Coastal Engineering*, London, UK, pp. 1141-1158.

Dingemans, M.W. (1997): *Water wave propagation over uneven bottoms. Part 2 –*

non-linear wave propagation. In Advanced Series on Ocean Engineering, No. 13, World Scientific, 967 pp.

Fenton, J. (2007): *Coast and harbour engineering*. Lecture Textbook, 80 pp. http://www.ifh.uni-karlsruhe.de/people/fenton/LectureNotes/Coastal-and-Ocean.pdf.

Galvin, C.J. (1968): Breaker type classification on three laboratory beaches. *Journal of Geophysical Research*, Vol. 73, No. 12, pp. 3651-3659.

Galvin, C.J. (1969): Breaker travel and choice of design wave height. *Journal of Waterways and Harbours*, No. 95, pp. 175-200.

Germain, J.P. (1984): *Coefficients de réflexion et de transmission en eau peu profonde*. Rozprawy Hydrotechniczne, Report No. 46, Instytut Budownictwa Wodnego, Gdańsk, Poland (in French), pp. 5-13.

Gobbi, M.F.; Kirby, J.T. (1999): Wave evolution over submerged sills: test of a high order Boussinesq model. *Coastal Engineering*, No. 37, pp. 57-96.

Gobbi, M.F.; Kirby, J.T.; Wei, G. (2000): A fully nonlinear Boussinesq model for surface waves. Part 2. Extension to $0(kh)^4$. *Journal of Fluid Mechanics*, No. 405, pp. 181-210.

Goda, Y. (1970): A synthesis of breaker indices. *Trans. Japanese Soc. Civil Engineers*, No. 2, Paper 2, pp. 227-230.

Goda, Y.; Okazaki, K.; Kagawa, M. (2000): Generation and evolution of harmonic wave components by abrupt depth changes. *Proceedings of Coastal Structures*, Rotterdam, Netherlands, pp. 649-658.

Goring, D.G. (1979): Tsunamis – the propagation of long waves onto a shelf. Ph.D. Thesis, California Institute of Technology Pasadena, USA, 305 pp.

Goring, D.G.; Raichlen, F. (1992): Propagation of long waves onto shelf. *Journal of Waterway, Port, Coastal, and Ocean Engineering*, Vol. 118, No. 1, pp. 43-61.

Grilli, S.T.; Skourup, J.; Svendsen, I.A. (1989): An efficient boundary element method for nonlinear water waves. *Engng. Anal. With Boundary Elements*, No. 6, pp. 97-107.

Grilli, S.T.; Losada, M.; Martin, F. (1994): Characteristics of solitary wave breaking induced by breakwaters. *Journal of Waterway, Port, Coastal, and Ocean Engineering,* Vol. 120, No. 1, pp. 74-92.

Grilli, S.T.; Giblert, R.W.; Libin, P.; Vincent, S.; Astruc, D.; Legendre, D.; Duval, M.; Kimmoun, O.; Branger, H.; Devrard, D.; Fraunie, P.; Abadie, S. (2004): Numerical modelling and experiments for solitary wave shoaling and breaking over a sloping beach. *Proceedings of the 14th International Offshore and Polar Engineering Conference*, Toulon, France, pp. 306-312.

Grüne, J.; Schmidt-Koppenhagen, R.; Oumeraci, H. (2006): Tsunami wave decay in near- and onshore areas. *Proceedings of the 30th International Conference on Coastal Engineering*, San Diego, USA, pp. 1664-1676.

Hara, M.; Yasuda, T.; Sakakibara, Y. (1992): Characteristics of a solitary wave breaking caused by a submerged obstacle. *Proceedings of the 23th International Conference on Coastal Engineering*, Venice, Italy, pp. 253-266.

Irribarren, C.R.; Nogales, C. (1949): Protection des ports II. *Proceedings of the 17th International Navig. Congress*, Lisbon, Portugal, pp. 31-80.

Isobe, M.; Watanabe, A.; Yamamoto, S. (1996): Nonlinear wave transformation due to a submerged breakwater. *Proceedings of the 25th International Conference on Coastal Engineering*, Orlando, USA, pp. 767-780.

Iwata, K.; Kawasaki, K.; Kim, D.S. (1996): Breaking limit, breaking and post – breaking wave deformation due to submerged structures. *Proceedings of the 25th International Conference on Coastal Engineering*, Orlando, USA, pp. 2338-2351.

Jeffreys, F.R.S. (1944): *Motion of waves in shallow water.* Wave Report No. 3, Great Britain Ministry of Supply, pp. 1-3.

Johnson, J.W.; Fuchs, R.A.; Morison, J.R. (1951): The damping action of submerged breakwaters. *Transactions American Geophysical Union*, Vol. 32, No. 5, pp. 704-718.

Johnson, R.S. (1972): Some numerical solutions of a variable-coefficient Korteweg-de-Vries equation (with applications to solitary wave development on a shelf). *Journal of Fluid Mechanics*, Vol. 54, No. 1, pp. 81-91.

Kabbaj, A. (1985): Contribution à l'étude du passaga des ondes de gravité et de la génération des ondes internes sur un talus, dans le cadre de la théorie de l'eau peu profonde. Thése, Université Scientifique et Médicale de Grenoble (in French).

Kennedy, A.B.; Chen, Q.; Kirby, J.T.; Dalrymple, R.A. (2000): Boussinesq modelling of wave transformation, breaking and runup. I: 1D. *Journal of Waterway, Port, Coastal, and Ocean Engineering*, Vol. 126, No. 1, pp. 39-47.

Khader, M.H.; Rai S.P. (1980): A study of submerged breakwaters. *Journal of Hydraulic Research*, Vol. 18, No. 2, pp. 113-121.

Kim, D.-H.; Lynett, P.J.; Socolofsky, S. (2009): A depth-integrated model for weakly dispersive, turbulent, and rotational fluid flows. *Ocean Modelling*, Vol. 27, No. 3-4, pp. 198-214.

Kirby, J.T.; Wei, G.; Chen, Q.; Kennedy, A.B.; Dalrymple, R.A. (1998): FUNWAVE 1.0. *Fully Nonlinear Boussinesq wave model documentation and user's manual*. Research Report NO. CACR-98-06, Center for Applied Coastal research, Department of Civil Engineering, University of Delaware, USA, 80 pp.

Kirby, J.T (2003): Boussinesq models and applications to nearshore wave propagation, surf zone processes and wave-induced currents. In Advances in Coastal Modelling, Elsevier Oceanography Series, pp. 1-41.

Kolmogorov, A.N. (1942): Equations of turbulent motion on an incompressible fluid. Izv. Akad. Naul. SSSR, Seria fizicheska Vi., pp. 56-58 (English translation: Imperial College, Mech. Eng. Dept. Rept. ON/6, 1968).

Korteweg, D.J.; de Vries, G. (1895): On the change of form of long waves advancing in a rectangular canal, and on a type of long stationary waves. *Phil. Mag. And J. of Science*, No. 39, pp. 422-443.

Koether, G.R. (2002): *Hydraulische Wirksamkeit getauchter Einzelfilter und Filtersysteme – Prozessbeschreibung und Modellbildung für ein innovatives Riffkonzept*. Ph.D. Thesis, Leichtweiß-Institute for Hydraulic Engineering and Water Resources, TU Braunschweig, Germany (in German), 218 pp.

Lamb, H. (1932). *Hydrodynamics*. Cambridge University Press, Cambridge.

Le Mehaute, B.; Divoky, D.; Lin, A. (1968): Shallow water waves: A comparison of theories and experiments. *Proceedings of the 11th International Conference on Coastal Engineering*, London, UK, pp. 86-107.

Lenau, C.W. (1966): The solitary wave of maximum amplitude. *Journal of Fluid Mechanics*, Vol. 26, Part II, pp. 309-320.

Lin, P. (2004): A numerical study of solitary wave interaction with rectangular obstacles. *Coastal Engineering*, No. 51, pp. 35-51.

Liu, P.L.-F. (1994): *Model equations for wave propagation from deep to shallow water*. In Advances in Coastal and Ocean Engineering, Vol. 1, World Scientific, pp. 125-157.

Liu, P.L.-F.; Cho, Y.S.; Yoon; S.B.; Seo, S.N. (1994): *Numerical simulations of the 1960 Chilean tsunami propagation and inundation at Hilo, Hawaii.* In Recent Developments in Tsunami Research, Kluwer Academic Publishers, pp. 99-115.

Liu, P.L.-F.; Lin, P. (1997): *A numerical model for breaking wave: the volume of fluid method.* Report CACR-97-02, Center for Applied Coastal Research, Ocean Eng. Lab., University of Delaware, USA, 54 pp.

Liu, P.L.-F.; Cheng, Y. (2001): A numerical study of the evolution of a solitary wave over a shelf. *Physics of Fluids*, Vol. 13, No. 6, pp. 1660-1667.

Liu, P.L.-F.; Losada, I.J. (2002): Wave propagation modelling in coastal engineering. *Journal of Hydraulic Research*, Vol. 40, No. 3, pp. 229-240.

Liu, P.L.-F. (2007): *Mathematical and numerical modelling of tsunamis in nearshore environment: present and future*. Presentation in the frame of DFG-Round Table Discussion: Near- and onshore tsunami effects. FZK, Hannover, Germany, http://www.fzk.uni-hannover.de/323.html.

Longuet-Higgins, M.S. (1974): On the mass, momentum, energy and circulation of a solitary wave. *Proc. Royal Society London A*, No. 337, pp. 1-13.

Longuet-Higgins, M.S.; Cokelet, E.D. (1976): The deformation of steep surface waves on water – I. A numerical method of computation. *Proc. Royal Society London A*, Vol. 350, pp. 1-26.

Losada, M.A.; Vidal, C.; Median, R. (1989): Experimental study on the evolution of a solitary wave at an abrupt junction. *Journal of Geophysical Research*, Vol. 94, No. C10, pp. 14557-14566.

Lynett, P.J. (2002): *A multi-layer approach to modelling generation, propagation, and interaction of water waves*. Ph.D. Thesis, Cornell University, USA, 201 pp. http://ceprofs.tamu.edu/plynett/cv/index.html.

Lynett, P.J.; Liu, P.L.-F. (2002): A Numerical Study of Submarine Landslide Generated Waves and Runup. *Proc. Royal Society of London A*, No. 458, pp. 2885-2910.

Lynett, P.J. (2006): Nearshore modelling with high-order Boussinesq equations. *Journal of Waterway, Port, Coastal, Ocean Engineering*, Vol. 132, No. 5, pp. 348-357.

Lynett, P.J. (2007): The effect of a shallow water obstruction on long wave runup and overland flow velocity. *Journal of Waterway, Port, Coastal, Ocean Engineering*, Vol. 133, No. 6, pp. 455-462.

Madsen, O.S.; Mei, C.C. (1969): The transformation of a solitary wave over an uneven bottom. *Journal of Fluid Mechanics*, Vol. 39, No. 4, pp. 781-791.

Madsen, P.A.; Murray, R.; Sørensen, O.R. (1991): A new form of the Boussinesq equations with improved linear dispersion characteristics. *Coastal Engineering*, No. 15, pp. 371-388.

Madsen, P.A.; Schäffer, H.A. (1998): Higher order Boussinesq equations for surface waves: derivation and analysis. *Phil. Trans. R. Soc. Lond. A*, No. 356, pp. 3123-3184.

Madsen, P.A.; Bingham, H.B.; Liu, H. (2002): A new Boussinesq method for fully nonlinear waves from shallow to deep water. *Journal of Fluid Mechanics*, No. 462. pp. 1-30.

Madsen, P.A.; Bingham, H.B.; Schäffer, H.A. (2003): Boussinesq-type formulations for fully nonlinear and extremely dispersive water waves. Derivation and analysis. *Proc. Roy. Soc. Lond. A*, No. 459, pp. 1075-1104.

Madsen, P.A.; Fuhrman, D.R.; Schäffer, H.A. (2008): On the solitary wave paradigm for tsunamis. *Journal of Geophysical Research - Oceans*, Vol. 113, pp. 1-22.

Matsuyama, M.; Ikeno, M.; Sakakiyama, T.; Takeda, T. (2007): A study on tsunami wave fission in an undistorted experiment. *Pure and Applied Geophysics*, No. 164, pp. 617-631.

McCowan, J. (1891): On the solitary wave. *Phil. Mag. J. Science*, Vol. 32, 5th Series, pp. 45-58.

McNair, E.C.; Sørensen, R.M. (1970): Characteristics of waves broken by a longshore bar. *Proceedings of the 12th International Conference on Coastal*

Engineering, Washington, USA, pp. 415-434.

Miche, M. (1944): Mouvements onulatoires de la mer en profondeur constante ou decroissante. *Annales de Ponts et Chaussees*.

Michell, J.H. (1893): On the highest waves in water. *Phil. Mag. J. Science*, No. 36, pp. 430-437.

Nakamura, M.; Shiraishi, H.; Sasaki, Y. (1966): Wave damping effect of submerged dike. *Proceedings of the 10th International Conference on Coastal Engineering*, Tokyo, Japan, pp. 254-267.

Nelson, R.C. (1987): Design wave heights on very mild slopes – An experimental study. *Civil Eng. Trans., Inst. Engs. Australia*, Vol. CE29, pp. 157-161.

Nwogu, O. (1993): Alternative form of Boussinesq equations for nearshore wave propagation. *Journal of Waterway, Port, Coastal, and Ocean Engineering*, No. 119, pp. 618-638.

Ohyama, T.; Nadaoka, K. (1994): Transformation of a nonlinear wave train passing over a submerged shelf without breaking. *Coastal Engineering*, No. 24, pp. 1-22.

Ohyama, T.; Kioka, W.; Tada, A. (1995): Applicability of numerical models to nonlinear dispersive waves. *Coastal Engineering*, No. 24, pp. 297-313.

Oumeraci, H. (2006): *Near- and onshore tsunami effects – knowledge base generation and model development*. Background Paper for DFG-Round Table Discussion 2007: Near- and onshore tsunami effects. FZK, Hannover, Germany, 19 pp. http://www.fzk.uni-hannover.de/323.html.

Patrick, D.A.; Wiegel, R.L. (1954): Amphibian tractors in the surf. *Proceedings of the Conference on Ships Waves*, 1, 397.

Peregrine, D.H. (1967): Long waves on a beach. *Journal of Fluid Mechanics*, No. 27, pp. 815-827.

Powell, K.A.; Allsop, N.W.H. (1985): *Low-crested breakwaters, hydraulic performance and stability*. Report SR 57, Hydraulic Research, Wallingford, UK, 23 pp.

Schäffer, H.A.; Madsen, P.A.; Deigaard, R. (1993): A Boussinesq model for waves breaking in shallow water. *Coastal Engineering*, No. 20, pp. 185-202.

Schüttrumpf, H. (2001): *Hydrodynamische Belastung der Binenböschung von Seedeichen durch Wellenüberlauf*. Mitteilungshefte des Leichtweiß-

Instituts, Heft 149, pp. 1-124 (in German).

Seabra-Santos, F.J.; Renouard, D.P.; Temperville, A.M. (1987): Numerical and experimental study of the transformation of a solitary wave over a shelf or isolated obstacle. *Journal of Fluid Mechanics*, No. 176, pp.117-134.

Seabrook, S.R.; Hall, K.R. (1998): Wave transformation at submerged rubble mound breakwaters. *Proceedings of the 26th International Conference on Coastal Engineering*, Copenhagen, Denmark, pp. 2000-2013.

Seelig, W.N. (1980): Two-dimensional tests of wave transmission and reflection characteristics of laboratory breakwaters. Technical Report No. 80-1, WES, CERC, Vicksburg, 187 pp.

Singamsetti, S.R.; Wind, H.G. (1980): Characteristics of shoaling and breaking periodic waves normally incident to plane beaches of constant slope. Delft Hydr. Lab. Report M1371.

Sitanggang, K.I.; Lynett, P.J.; Liu, P.L.-F. (2006): Development of a Boussinesq-RANS VOF hybrid wave model. *Proceedings of the 30th International Conference on Coastal Engineering*, San Diego, USA, pp. 24-35.

Smagorinsky, J.S (1963): General circulation experiments with the primitive equations. *Mon. Weather Rev.*, No. 93, pp. 99-.

Smith, E.R.; Kraus, N.C. (1991): *Laboratory study on macro-features of wave breaking over bars and artificial reefs*. Report CERC-90-12, Coastal Engineering Research Center, Department of The Army, Vicksburg, USA, 232 pp.

Strusińska, A.; Oumeraci, H. (2008): Application of artificial reef to tsunami hazard mitigation: laboratory investigation on tsunami nonlinear transformation at a submerged structure of finite width. *Proceedings of the International Conference on Tsunami Warning*, Bali, Indonesia.

Strusińska, A.; Oumeraci, H. (2009a): *Hydraulic performance of artificial reef for tsunami protection: state of the art. Part I.* Internal Report, Leichtweiß-Institute for Hydraulic Engineering and Water Resources, TU Braunschweig, Germany, 138 pp.

Strusińska, A.; Oumeraci, H. (2009b): *Hydraulic performance of artificial reef for tsunami protection: state of the art. Part II.* Internal Report, Leichtweiß-Institute for Hydraulic Engineering and Water Resources, TU

Braunschweig, Germany, 200 pp.

Strusińska, A.; Oumeraci, H. (2009c): *Hydraulic performance of artificial reef for tsunami protection: description and validation of numerical model COULWAVE. Part I.* Internal Report, Leichtweiß-Institute for Hydraulic Engineering and Water Resources, TU Braunschweig, Germany, 179 pp.

Strusińska, A.; Oumeraci, H. (2009d): *Hydraulic performance of artificial reef for tsunami protection: description and validation of numerical model COULWAVE. Part II.* Internal Report, Leichtweiß-Institute for Hydraulic Engineering and Water Resources, TU Braunschweig, Germany, 58 pp.

Strusińska, A.; Oumeraci, H. (2009e): *Hydraulic performance of artificial reef for tsunami protection: laboratory experiments on solitary wave nonlinear transformation over a submerged structure of finite width.* Internal Report, Leichtweiß-Institute for Hydraulic Engineering and Water Resources, TU Braunschweig, Germany, 241 pp.

Sue, Y.C.; Chern, M.J.; Hwang, R.R. (2005): Interaction of nonlinear progressive viscous waves with a submerged obstacle. *Ocean Engineering*, No. 32, pp. 893-923.

Svendsen, I.A. (2006): *Introduction to nearshore hydrodynamics*. Advanced Series on Ocean Engineering, World Scientific, 722 pp.

Tappert, F.D.; Zabusky, N.J. (1971): Gradient-induced fission of solitons. *Physical Review Letters*, Vol. 27, No. 26, pp. 1774-1776.

Ting, C.L.; Lin, M.C.; Hsu, C.M (2005): Spatial variations of waves propagating over a submerged rectangular obstacle. *Ocean Engineering*, No. 32, pp. 1448-1464.

Titov, V.V.; Synolakis, C.E. (1998): Numerical modeling of tidal wave runup. *Journal of Waterway, Port, Coastal and Ocean Engineering*, Vol. 124, No. 4, pp. 157–171.

Ursell, F. (1953): The long wave paradox. *Proceedings of the Cambridge Philosophical Society*, Vol. 49, pp. 685-694.

Van der Meer, J.W.; d`Angremond, K. (1991): Wave transmission at low-crested structures. *Proceedings of the Conference on Coastal Structures and Breakwaters*, London, UK, pp. 4-25.

Ward, S.N. (2004): *Tsunamis*. Encyclopedia of Physical Science and Technology,

Academic Press, pp. 16.

Weggel, J.R. (1972): Maximum breakwater height. *Journal of Waterways, Harbors and Coastal Engineering Division*, Vol. 98, pp. 529-548.

Wei, G.; Kirby, J.T. (1995): Time-dependent numerical code for extended Boussinesq equations. *Journal of Waterway, Port, Coastal, and Ocean Engineering*, Vol. 121, No. 5, pp. 251-261.

Wei, G.; Kirby, J.T.; Grilli, S.T.; Subramanya, R. (1995): A fully nonlinear Boussinesq model for surface waves. Part I. Highly nonlinear unsteady waves. *Journal of Fluid Mechanics*, No. 7, pp. 273-286.

Witting, G.B. (1984): A unified model for the evolution of nonlinear water waves. *Journal of Computational Physics*, Vol. 56, pp. 203-236.

Yasuda, T. (2007): Tsunami pressure acting upon a vertical wall on a beach. *Presentation in the framework of the Coastal Structure Conference*, Venice, Italy.

Yoo, D. (1986): *Mathematical modelling of wave-current interacted flow in shallow waters*. Ph.D. Thesis, University of Manchester, UK.

Yoon, S.B., Liu, P.L.-F. (1989): Interaction of currents and weakly nonlinear waves in shallow water. *Journal of Fluid Mechanics*, No. 205, pp. 397-419.

Yoshioka, K.; Kawakami, T.; Tanaka, S.; Koarai, S.; Uda, T. (1993): Design Manual for Artificial Reefs. *Coastlines of Japan II*, Coastal Zone, ASCE, pp. 93-107.

Zelt, J.A. (1991): The runup of nonbreaking and breaking solitary waves. *Coastal Engineering*, No. 15, pp. 205-246.

Appendix A

EXPERIMENTAL PROGRAMME FOR 2007 AND 2008 LABORATORY TESTS IN THE LWI WAVE FLUME

Table A1: Programme of the laboratory experiments performed in 2007

TEST NO.	WAVE TYPE	$H_{i,nom}$ [m]	h [m]	h_r [m]	d_r [m]	B [m]	REEF SLOPES	AIM
1	Solitary	0.06	0.6	0.3	0.3	2.0	1:2	WB, WF
2	Solitary	0.12	0.6	0.3	0.3	2.0	1:2	WB, WF
3	Solitary	0.18	0.6	0.3	0.3	2.0	1:2	WB, WF
4	Solitary	0.22	0.6	0.3	0.3	2.0	1:2	WB, WF
5	Solitary	0.06	0.6	0.3	0.3	2.0	No	WB, WF
6	Solitary	0.12	0.6	0.3	0.3	2.0	No	WB, WF
7	Solitary	0.18	0.6	0.3	0.3	2.0	No	WB, WF
8	Solitary	0.22	0.6	0.3	0.3	2.0	No	WB, WF
9	Solitary	0.06	0.6	0.4	0.2	2.0	1:2	WB, WF
10	Solitary	0.12	0.6	0.4	0.2	2.0	1:2	WB, WF
11	Solitary	0.18	0.6	0.4	0.2	2.0	1:2	WB, WF
12	Solitary	0.22	0.6	0.4	0.2	2.0	1:2	WB, WF
13	Solitary	0.06	0.6	0.4	0.2	2.0	No	WB, WF
14	Solitary	0.12	0.6	0.4	0.2	2.0	No	WB, WF
15	Solitary	0.18	0.6	0.4	0.2	2.0	No	WB, WF
16	Solitary	0.22	0.6	0.4	0.2	2.0	No	WB, WF
17	Solitary	0.06	0.6	0.5	0.1	2.0	1:2	WB, WF
18	Solitary	0.12	0.6	0.5	0.1	2.0	1:2	WB, WF
19	Solitary	0.18	0.6	0.5	0.1	2.0	1:2	WB, WF
20	Solitary	0.22	0.6	0.5	0.1	2.0	1:2	WB, WF
21	Solitary	0.06	0.6	0.4	0.1	2.0	No	WB, WF
22	Solitary	0.12	0.6	0.4	0.1	2.0	No	WB, WF
23	Solitary	0.18	0.6	0.4	0.1	2.0	No	WB, WF
24	Solitary	0.22	0.6	0.4	0.1	2.0	No	WB, WF
25	Solitary	0.06	0.6	0.5	0.1	1.0	1:2	WB, WF
26	Solitary	0.12	0.6	0.5	0.1	1.0	1:2	WB, WF
27	Solitary	0.18	0.6	0.5	0.1	1.0	1:2	WB, WF
28	Solitary	0.22	0.6	0.5	0.1	1.0	1:2	WB, WF
29	Solitary	0.06	0.6	0.5	0.1	1.0	No	WB, WF
30	Solitary	0.12	0.6	0.5	0.1	1.0	No	WB, WF
31	Solitary	0.18	0.6	0.5	0.1	1.0	No	WB, WF
32	Solitary	0.22	0.6	0.5	0.1	1.0	No	WB, WF

Table A1: Continuation: Programme of the laboratory experiments performed in 2007

TEST NO.	WAVE TYPE	$H_{i,nom}$ [m]	h [m]	h_r [m]	d_r [m]	B [m]	REEF SLOPES	AIM
33	Solitary	0.06	0.6	0.4	0.2	1.0	1:2	WB, WF
34	Solitary	0.12	0.6	0.4	0.2	1.0	1:2	WB, WF
35	Solitary	0.18	0.6	0.4	0.2	1.0	1:2	WB, WF
36	Solitary	0.22	0.6	0.4	0.2	1.0	1:2	WB, WF
37	Solitary	0.06	0.6	0.4	0.2	1.0	No	WB, WF
38	Solitary	0.12	0.6	0.4	0.2	1.0	No	WB, WF
39	Solitary	0.18	0.6	0.4	0.2	1.0	No	WB, WF
40	Solitary	0.22	0.6	0.4	0.2	1.0	No	WB, WF
41	Solitary	0.06	0.6	0.3	0.1	1.0	1:2	WB, WF
42	Solitary	0.12	0.6	0.3	0.1	1.0	1:2	WB, WF
43	Solitary	0.18	0.6	0.3	0.1	1.0	1:2	WB, WF
44	Solitary	0.22	0.6	0.3	0.1	1.0	1:2	WB, WF
45	Solitary	0.06	0.6	0.3	0.1	1.0	No	WB, WF
46	Solitary	0.12	0.6	0.3	0.1	1.0	No	WB, WF
47	Solitary	0.18	0.6	0.3	0.1	1.0	No	WB, WF
48	Solitary	0.22	0.6	0.3	0.1	1.0	No	WB, WF
49	Solitary	0.06	0.6	0.0	0.6	0.0	No	WB, WF
50	Solitary	0.12	0.6	0.0	0.6	0.0	No	WB, WF
51	Solitary	0.18	0.6	0.0	0.6	0.0	No	WB, WF
52	Solitary	0.22	0.6	0.0	0.6	0.0	No	WB, WF

WB – wave breaking, WF – wave fission

Table A2: Programme of the laboratory experiments performed in 2008

TEST NO.	WAVE TYPE	H_i [m]	h [m]	h_r [m]	d_r [m]	B [m]	REEF SLOPES	AIM
1	Solitary	0.08	0.6	0.3	0.3	1.0	1:2	WF
2	Solitary	0.10	0.6	0.3	0.3	1.0	1:2	WF
3	Solitary	0.12	0.6	0.3	0.3	1.0	1:2	WF
4	Solitary	0.14	0.6	0.3	0.3	1.0	1:2	WF
5	Solitary	0.16	0.6	0.3	0.3	1.0	1:2	WF
6	Solitary	0.18	0.6	0.3	0.3	1.0	1:2	WF
7	Solitary	0.20	0.6	0.3	0.3	1.0	1:2	WB, WF
8	Solitary	0.12	0.5	0.3	0.2	1.0	1:2	WB
9	Solitary	0.14	0.5	0.3	0.2	1.0	1:2	WB
10	Solitary	0.16	0.5	0.3	0.2	1.0	1:2	WB
11	Solitary	0.18	0.5	0.3	0.2	1.0	1:2	WB
12	Solitary	0.20	0.5	0.3	0.2	1.0	1:2	WB
13	Solitary	0.22	0.5	0.3	0.2	1.0	1:2	WB
14	Solitary	0.08	0.6	0.4	0.2	1.0	1:2	WF
15	Solitary	0.10	0.6	0.4	0.2	1.0	1:2	WF
16	Solitary	0.14	0.6	0.4	0.2	1.0	1:2	WB, WF
17	Solitary	0.16	0.6	0.4	0.2	1.0	1:2	WB, WF
18	Solitary	0.18	0.6	0.4	0.2	1.0	1:2	WB
19	Solitary	0.20	0.6	0.4	0.2	1.0	1:2	WB
20	Solitary	0.22	0.7	0.4	0.3	1.0	1:2	WB
21	Solitary	0.08	0.5	0.4	0.1	1.0	1:2	WB
22	Solitary	0.10	0.5	0.4	0.1	1.0	1:2	WB
23	Solitary	0.12	0.5	0.4	0.1	1.0	1:2	WB
24	Solitary	0.14	0.5	0.4	0.1	1.0	1:2	WB
25	Solitary	0.16	0.5	0.4	0.1	1.0	1:2	WB
26	Solitary	0.18	0.5	0.4	0.1	1.0	1:2	WB
27	Solitary	0.20	0.5	0.4	0.1	1.0	1:2	WB
28	Solitary	0.22	0.5	0.4	0.1	1.0	1:2	WB
29	Solitary	0.18	0.7	0.5	0.2	1.0	1:2	WB
30	Solitary	0.20	0.7	0.5	0.2	1.0	1:2	WB
31	Solitary	0.22	0.7	0.5	0.2	1.0	1:2	WB

Table A2: Continuation: Programme of the laboratory experiments performed in 2008

TEST NO.	WAVE TYPE	H_i [m]	h [m]	h_r [m]	d_r [m]	B [m]	REEF SLOPES	AIM
32	Solitary	0.06	0.6	0.5	0.1	1.0	1:2	WF
33	Solitary	0.08	0.6	0.5	0.1	1.0	1:2	WF
34	Solitary	0.10	0.6	0.5	0.1	1.0	1:2	WB, WF
35	Solitary	0.12	0.6	0.5	0.1	1.0	1:2	WF
36	Solitary	0.14	0.6	0.5	0.1	1.0	1:2	WB, WF
37	Solitary	0.16	0.6	0.5	0.1	1.0	1:2	WB, WF
38	Solitary	0.20	0.6	0.5	0.1	1.0	1:2	WB
39	Solitary	0.10	0.7	0.5	0.2	2.0	1:2	WB
40	Solitary	0.12	0.7	0.5	0.2	2.0	1:2	WB
41	Solitary	0.14	0.7	0.5	0.2	2.0	1:2	WB
42	Solitary	0.16	0.7	0.5	0.2	2.0	1:2	WB
43	Solitary	0.18	0.7	0.5	0.2	2.0	1:2	WB
44	Solitary	0.20	0.7	0.5	0.2	2.0	1:2	WB
45	Solitary	0.22	0.7	0.5	0.2	2.0	1:2	WB
46	Solitary	0.06	0.6	0.5	0.1	2.0	1:2	WF
47	Solitary	0.08	0.6	0.5	0.1	2.0	1:2	WB
48	Solitary	0.10	0.6	0.5	0.1	2.0	1:2	WB
49	Solitary	0.14	0.6	0.5	0.1	2.0	1:2	WB
50	Solitary	0.16	0.6	0.5	0.1	2.0	1:2	WB
51	Solitary	0.20	0.6	0.5	0.1	2.0	1:2	WB
52	Solitary	0.14	0.7	0.4	0.3	2.0	1:2	WB
53	Solitary	0.16	0.7	0.4	0.3	2.0	1:2	WB
54	Solitary	0.18	0.7	0.4	0.3	2.0	1:2	WB
55	Solitary	0.20	0.7	0.4	0.3	2.0	1:2	WB
56	Solitary	0.22	0.7	0.4	0.3	2.0	1:2	WB
57	Solitary	0.06	0.5	0.4	0.1	2.0	1:2	WB
58	Solitary	0.08	0.5	0.4	0.1	2.0	1:2	WB
59	Solitary	0.10	0.5	0.4	0.1	2.0	1:2	WB
60	Solitary	0.12	0.5	0.4	0.1	2.0	1:2	WB
61	Solitary	0.14	0.5	0.4	0.1	2.0	1:2	WB

Table A2: Continuation: Programme of the laboratory experiments performed in 2008

TEST NO.	WAVE TYPE	H_i [m]	h [m]	h_r [m]	d_r [m]	B [m]	REEF SLOPES	AIM
62	Solitary	0.06	0.6	0.4	0.2	2.0	1:2	WF
63	Solitary	0.08	0.6	0.4	0.2	2.0	1:2	WF
64	Solitary	0.10	0.6	0.4	0.2	2.0	1:2	WB, WF
65	Solitary	0.14	0.6	0.4	0.2	2.0	1:2	WB, WF
66	Solitary	0.16	0.6	0.4	0.2	2.0	1:2	WB, WF
67	Solitary	0.20	0.6	0.4	0.2	2.0	1:2	WB
68	Solitary	0.18	0.7	0.3	0.4	2.0	1:2	WB
69	Solitary	0.20	0.7	0.3	0.4	2.0	1:2	WB
70	Solitary	0.22	0.7	0.3	0.4	2.0	1:2	WB
71	Solitary	0.10	0.5	0.3	0.2	2.0	1:2	WB
72	Solitary	0.12	0.5	0.3	0.2	2.0	1:2	WB
73	Solitary	0.14	0.5	0.3	0.2	2.0	1:2	WB
74	Solitary	0.08	0.6	0.3	0.3	2.0	1:2	WF
75	Solitary	0.10	0.6	0.3	0.3	2.0	1:2	WF
76	Solitary	0.14	0.6	0.3	0.3	2.0	1:2	WB, WF
77	Solitary	0.16	0.6	0.3	0.3	2.0	1:2	WB, WF
78	Solitary	0.20	0.6	0.3	0.3	2.0	1:2	WB, WF
79	Solitary	0.22	0.6	0.3	0.3	2.0	1:2	WF

WB – wave breaking, WF – wave fission

Appendix B

REEF GEOMETRIES AND INCIDENT SOLITARY WAVE CHARACTERISTICS IN EXPERIMENTS 2007 AND 2008 FOR MEASURED INCIDENT WAVE HEIGHTS $H_{i,1}$, $H_{i,2}$, $H_{i,3}$

Table B1: Incident solitary wave conditions for different definitions of incident wave heights for water depth h=0.5 m in 2007 and 2008 experiments

	RELATIVE INCIDENT WAVE HEIGHT	RELATIVE INCIDENT WAVE HEIGHT H_i/d_r [-]		RELATIVE STRUCTURE WIDTH B/L_i [-]	
	H_i/h [-]	d_r=0.2m	d_r=0.1m	B=1.0m	B=2.0m
Incident wave height $H_{i,1}$					
H_i=0.060 m L_i=6.12 m	0.12	0.30	0.60	0.16	0.33
H_i=0.082 m L_i=5.23 m	0.16	0.41	0.82	0.19	0.38
H_i=0.103 m L_i=4.67 m	0.21	0.52	1.03	0.21	0.43
H_i=0.122 m L_i=4.29 m	0.24	0.61	1.22	0.23	0.47
H_i=0.143 m L_i=3.96 m	0.29	0.72	1.43	0.25	0.51
H_i=0.166 m L_i=3.68 m	0.33	0.83	1.66	0.27	0.54
H_i=0.180 m L_i=3.53 m	0.36	0.90	1.80	0.28	0.57
H_i=0.205 m L_i=3.31 m	0.41	1.03	2.05	0.30	0.60
H_i=0.224 m L_i=3.17 m	0.45	1.12	2.24	0.32	0.63
Incident wave height $H_{i,2}$					
H_i=0.060 m L_i=6.12 m	0.12	0.30	0.60	0.16	0.33
H_i=0.085 m L_i=5.14 m	0.17	0.43	0.85	0.19	0.39
H_i=0.106 m L_i=4.60 m	0.21	0.53	1.06	0.22	0.43
H_i=0.127 m L_i=4.21 m	0.25	0.64	1.27	0.24	0.48
H_i=0.152 m L_i=3.85 m	0.30	0.76	1.52	0.26	0.52
H_i=0.176 m L_i=3.57 m	0.35	0.88	1.76	0.28	0.56
H_i=0.191 m L_i=3.43 m	0.38	0.96	1.91	0.29	0.58
H_i=0.219 m L_i=3.20 m	0.44	1.10	2.19	0.31	0.63
H_i=0.237 m L_i=3.08 m	0.47	1.19	2.37	0.32	0.65

Table B1: Continuation: Incident solitary wave conditions for different definitions of incident wave heights for water depth h=0.5 m in 2007 and 2008 experiments

	RELATIVE INCIDENT WAVE HEIGHT H_i/h [-]	**RELATIVE INCIDENT WAVE HEIGHT H_i/d_r [-]**		**RELATIVE STRUCTURE WIDTH B/L_i [-]**	
		d_r=0.2m	**d_r=0.1m**	**B=1.0m**	**B=2.0m**
Incident wave height $H_{i,3}$					
H_i=0.061 m L_i=6.07 m	0.12	0.31	0.61	0.16	0.33
H_i=0.089 m L_i=5.02 m	0.18	0.45	0.89	0.20	0.40
H_i=0.110 m L_i=4.52 m	0.22	0.55	1.10	0.22	0.44
H_i=0.133 m L_i=4.11 m	0.27	0.67	1.33	0.24	0.49
H_i=0.161 m L_i=3.74 m	0.32	0.81	1.61	0.27	0.53
H_i=0.187 m L_i=3.47 m	0.37	0.94	1.87	0.29	0.58
H_i=0.201 m L_i=3.34 m	0.40	1.01	2.01	0.30	0.60
H_i=0.234 m L_i=3.10 m	0.47	1.17	2.34	0.32	0.65
H_i=0.251 m L_i=2.99 m	0.50	1.26	2.51	0.33	0.67

Table B2: Incident solitary wave conditions for different definitions of incident wave heights for water depth h=0.6 m in 2007 and 2008 experiments

	RELATIVE INCIDENT WAVE HEIGHT	RELATIVE INCIDENT WAVE HEIGHT H_i/d_r [-]			RELATIVE STRUCTURE WIDTH B/L_i [-]	
	H_i/h [-]	d_r=0.3m	d_r=0.2m	d_r=0.1m	B=1.0m	B=2.0m
Incident wave height $H_{i,1}$						
H_i=0.060 m L_i=8.04 m	0.10	0.20	0.30	0.60	0.12	0.25
H_i=0.082 m L_i=6.88 m	0.14	0.27	0.41	0.82	0.15	0.29
H_i=0.103 m L_i=6.14 m	0.17	0.34	0.52	1.03	0.16	0.33
H_i=0.122 m L_i=5.64 m	0.20	0.41	0.61	1.22	0.18	0.35
H_i=0.143 m L_i=5.21 m	0.24	0.48	0.72	1.43	0.19	0.38
H_i=0.166 m L_i=4.84 m	0.28	0.55	0.83	1.66	0.21	0.41
H_i=0.180 m L_i=4.64 m	0.30	0.60	0.90	1.80	0.22	0.43
H_i=0.205 m L_i=4.35 m	0.34	0.68	1.03	2.05	0.23	0.46
H_i=0.224 m L_i=4.16 m	0.37	0.75	1.12	2.24	0.24	0.48
Incident wave height $H_{i,2}$						
H_i=0.060 m L_i=8.04 m	0.10	0.20	0.30	0.60	0.12	0.25
H_i=0.085 m L_i=6.76 m	0.14	0.28	0.43	0.85	0.15	0.30
H_i=0.106 m L_i=6.05 m	0.18	0.35	0.53	1.06	0.17	0.33
H_i=0.127 m L_i=5.53 m	0.21	0.42	0.64	1.27	0.18	0.36
H_i=0.152 m L_i=5.05 m	0.25	0.51	0.76	1.52	0.20	0.40
H_i=0.176 m L_i=4.70 m	0.29	0.59	0.88	1.76	0.21	0.43
H_i=0.191 m L_i=4.51 m	0.32	0.64	0.96	1.91	0.22	0.44
H_i=0.219 m L_i=4.21 m	0.37	0.73	1.10	2.19	0.24	0.48
H_i=0.237 m L_i=4.05 m	0.40	0.79	1.19	2.37	0.25	0.49

Table B2: Continuation: Incident solitary wave conditions for different definitions of incident wave heights for water depth h=0.6 m in 2007 and 2008 experiments

	RELATIVE INCIDENT WAVE HEIGHT H_i/h [-]	**RELATIVE INCIDENT WAVE HEIGHT H_i/d_r [-]**			**RELATIVE STRUCTURE WIDTH B/L_i [-]**	
		d_r=0.3m	**d_r=0.2m**	**d_r=0.1m**	**B=1.0m**	**B=2.0m**
Incident wave height $H_{i,3}$						
H_i=0.061 m L_i=7.98 m	0.10	0.20	0.31	0.61	0.13	0.25
H_i=0.089 m L_i=6.61 m	0.15	0.30	0.45	0.89	0.15	0.30
H_i=0.110 m L_i=5.94 m	0.18	0.37	0.55	1.10	0.17	0.34
H_i=0.133 m L_i=5.40 m	0.22	0.44	0.67	1.33	0.19	0.37
H_i=0.161 m L_i=4.91 m	0.27	0.54	0.81	1.61	0.20	0.41
H_i=0.187 m L_i=4.56 m	0.31	0.62	0.94	1.87	0.22	0.44
H_i=0.201 m L_i=4.40 m	0.34	0.67	1.01	2.01	0.23	0.45
H_i=0.234 m L_i=4.07 m	0.39	0.78	1.17	2.34	0.25	0.49
H_i=0.251 m L_i=3.93 m	0.42	0.84	1.26	2.51	0.25	0.51

Table B3: Incident solitary wave conditions for different definitions of incident wave heights for water depth h=0.7 m in 2007 and 2008 experiments

	RELATIVE INCIDENT WAVE HEIGHT H_i/h [-]	RELATIVE INCIDENT WAVE HEIGHT H_i/d_r [-]			RELATIVE STRUCTURE WIDTH B/L_i [-]	
		d_r=0.4m	d_r=0.3m	d_r=0.2m	B=1.0m	B=2.0m
Incident wave height $H_{i,1}$						
H_i=0.103 m L_i=7.74 m	0.15	0.26	0.34	0.52	0.13	0.26
H_i=0.122 m L_i=7.11 m	0.17	0.31	0.41	0.61	0.14	0.28
H_i=0.143 m L_i=6.57 m	0.20	0.36	0.48	0.72	0.15	0.30
H_i=0.166 m L_i=6.09 m	0.24	0.42	0.55	0.83	0.16	0.33
H_i=0.18 m L_i=5.85 m	0.26	0.45	0.60	0.90	0.17	0.34
H_i=0.205 m L_i=5.48 m	0.29	0.51	0.68	1.03	0.18	0.36
H_i=0.224 m L_i=5.25 m	0.32	0.56	0.75	1.12	0.19	0.38
Incident wave height $H_{i,2}$						
H_i=0.106 m L_i=7.63 m	0.15	0.27	0.35	0.53	0.13	0.26
H_i=0.127 m L_i=6.97 m	0.18	0.32	0.42	0.64	0.14	0.29
H_i=0.152 m L_i=6.37 m	0.22	0.38	0.51	0.76	0.16	0.31
H_i=0.176 m L_i=5.92 m	0.25	0.44	0.59	0.88	0.17	0.34
H_i=0.191 m L_i=5.68 m	0.27	0.48	0.64	0.96	0.18	0.35
H_i=0.219 m L_i=5.31 m	0.31	0.55	0.73	1.10	0.19	0.38
H_i=0.237 m L_i=5.10 m	0.34	0.59	0.79	1.19	0.20	0.39
Incident wave height $H_{i,3}$						
H_i=0.110 m L_i=7.49 m	0.16	0.28	0.37	0.55	0.13	0.27
H_i=0.133 m L_i=6.81 m	0.19	0.33	0.44	0.67	0.15	0.29
H_i=0.161 m L_i=6.19 m	0.23	0.40	0.54	0.81	0.16	0.32
H_i=0.187 m L_i=5.74 m	0.27	0.47	0.62	0.94	0.17	0.35
H_i=0.201 m L_i=5.54 m	0.29	0.50	0.67	1.01	0.18	0.36
H_i=0.234 m L_i=5.13 m	0.33	0.59	0.78	1.17	0.19	0.39
H_i=0.251 m L_i=4.96 m	0.36	0.63	0.84	1.26	0.20	0.40

Appendix C

WAVE BREAKING CHARACTERISTICS IN LABORATORY EXPERIMENTS 2007 AND 2008

Table C1: Wave breaking characteristics for trapezoidal and rectangular structures of width **B=1.0 m** in 2007 (preliminary) experiments (H_i=$H_{i,nom}$)

				GENERATION OF WAVE BREAKING	LOCATION OF WAVE BREAKING		BREAKER TYPE	INTENSITY OF ENERGY DISSIPATION
					WAVE GAUGE	WAVE GAUGE LOCATION [m]		
B=1.0m	d_r/h=0.50	with reef slopes 1:2	H_i/h=0.10	No	-	-	-	-
			H_i/h=0.20	No	-	-	-	-
			H_i/h=0.30	No	-	-	-	-
			H_i/h=0.37	No	-	-	-	-
		without reef slopes	H_i/h=0.10	No	-	-	-	-
			H_i/h=0.20	No	-	-	-	-
			H_i/h=0.30	No	-	-	-	-
			H_i/h=0.37	No	-	-	-	-
	d_r/h=0.33	with reef slopes 1:2	H_i/h=0.10	No	-	-	-	-
			H_i/h=0.20	No	-	-	-	-
			H_i/h=0.30	No	-	-	-	-
			H_i/h=0.37	No	-	-	-	-
		without reef slopes	H_i/h=0.10	No	-	-	-	-
			H_i/h=0.20	No	-	-	-	-
			H_i/h=0.30	No	-	-	-	-
			H_i/h=0.37	No	-	-	-	-
	d_r/h=0.17	with reef slopes 1:2	H_i/h=0.10	No	-	-	-	-
			H_i/h=0.20	No	-	-	-	-
			H_i/h=0.30	Yes	WG9 ^{v}	37.37 BR	Transition	Medium
			H_i/h=0.37	Yes	WG9 ^{v,m}	36.77 LRT	Plunging	Medium
		without reef slopes	H_i/h=0.10	No	-	-	-	-
			H_i/h=0.20	No	-	-	-	-
			H_i/h=0.30	Yes	WG9 ^{v}	36.90 BR	Spilling	Weak
			H_i/h=0.37	Yes	WG9 ^{v}	36.90 BR	Plunging	Weak

BR – behind reef, LRT – landward reef toe, ^{m} – defined from measurements, ^{v} – defined from video records

Table C2: Wave breaking characteristics for trapezoidal and rectangular structures of width **B=2.0 m** in 2007 (preliminary) experiments ($H_i=H_{i,nom}$)

				GENERATION OF WAVE BREAKING	LOCATION OF WAVE BREAKING		BREAKER TYPE	INTENSITY OF ENERGY DISSIPATION
					WAVE GAUGE	WAVE GAUGE LOCATION [m]		
B=2.0m	$d_r/h=0.50$	with reef slopes 1:2	$H_i/h=0.10$	No	-	-	-	-
			$H_i/h=0.20$	No	-	-	-	-
			$H_i/h=0.30$	No	-	-	-	-
			$H_i/h=0.37$	No	-	-	-	-
		without reef slopes	$H_i/h=0.10$	No	-	-	-	-
			$H_i/h=0.20$	No	-	-	-	-
			$H_i/h=0.30$	No	-	-	-	-
			$H_i/h=0.37$	No	-	-	-	-
	$d_r/h=0.33$	with reef slopes 1:2	$H_i/h=0.10$	No	-	-	-	-
			$H_i/h=0.20$	No	-	-	-	-
			$H_i/h=0.30$	Yes	WG11 [v, m]	38.09 BR	Spilling	Medium
			$H_i/h=0.37$	Yes	WG10 [v]	37.59 LRT	Plunging	Medium
		without reef slopes	$H_i/h=0.10$	No	-	-	-	-
			$H_i/h=0.20$	No	-	-	-	-
			$H_i/h=0.30$	Yes	WG11 [v, m]	37.93 BR	Spilling	Weak
			$H_i/h=0.37$	Yes	WG10 [v]	37.68 BR	Plunging	Weak
	$d_r/h=0.17$	with reef slopes 1:2	$H_i/h=0.10$	No	-	-	-	-
			$H_i/h=0.20$	Yes	WG10 [v, m]	37.54 LRS	Plunging	Weak
			$H_i/h=0.30$	Yes	WG8 [m]	36.72 LRC	Plunging	Medium
			$H_i/h=0.37$	Yes	WG8 [v, m]	36.72 LRC	Plunging	Strong
		without reef slopes	$H_i/h=0.10$	No	-	-	-	-
			$H_i/h=0.20$	Yes	WG10 [v, m]	37.54 BR	Transition	Weak
			$H_i/h=0.30$	Yes	WG8 [m]	36.79 LRC	Plunging	Medium
			$H_i/h=0.37$	Yes	WG8 [v, m]	36.79 LRC	Plunging	Medium

BR – behind reef, LRC – landward reef corner, LRT– landward reef toe, LRS – landward reef slope, [m] – defined from measurements, [v] – defined from video records

Table C3: Detailed characteristics of solitary wave breaking for trapezoidal reef of width **B=1.0 m** in 2007 and 2008 experiments (H_i=$H_{i,nom}$)

				GENERATION OF WAVE BREAKING	LOCATION OF WAVE BREAKING		BREAKER TYPE	INTENSITY OF ENERGY DISSIPATION
					WAVE GAUGE	WAVE GAUGE LOCATION [m]		
B=1.0m, reef slopes 1:2	h=0.5m	d_r/h=0.4	H_i/h=0.24	No	-	-	-	-
			H_i/h=0.28	No	-	-	-	-
			H_i/h=0.32	No	-	-	-	-
			H_i/h=0.36	No	-	-	-	-
			H_i/h=0.40	No	-	-	-	-
			H_i/h=0.44	Yes	WG12 ^{v,m}	40.29 BR	Spilling	Weak
		d_r/h=0.2	H_i/h=0.16	No	-	-	-	-
			H_i/h=0.20	No	-	-	-	-
			H_i/h=0.24	No	-	-	-	-
			H_i/h=0.28	Yes	WG12 ^{m}	37.12 BR	Spilling	Very weak
			H_i/h=0.32	Yes	WG12 ^{v,m}	37.12 BR	Spilling	Medium
			H_i/h=0.36	Yes	WG22-8 ^{v}	36.40-36.60 LRS-LRT	Plunging	Weak
			H_i/h=0.40	Yes	WG7-22 ^{v}	36.20-36.40 LRS-LRS	Plunging	Medium
			H_i/h=0.44	Yes	WG7-22 ^{v}	36.20-36.40 LRS-LRS	Plunging	Medium
	h=0.6m	d_r/h=0.5	H_i/h=0.10	No	-	-	-	-
			H_i/h=0.20	No	-	-	-	-
			H_i/h=0.30	No	-	-	-	-
			H_i/h=0.33	No	-	-	-	-
			H_i/h=0.37	No	-	-	-	-
		d_r/h=0.33	H_i/h=0.10	No	-	-	-	-
			H_i/h=0.20	No	-	-	-	-
			H_i/h=0.23	No	-	-	-	-
			H_i/h=0.27	No	-	-	-	-
			H_i/h=0.30	No	-	-	-	-
			H_i/h=0.33	No	-	-	-	-
			H_i/h=0.37	No	-	-	-	-

175

Table C3: Continuation: Detailed characteristics of solitary wave breaking for trapezoidal reef of width **B=1.0 m** in 2007 and 2008 experiments ($H_i=H_{i,nom}$)

				GENERATION OF WAVE BREAKING	LOCATION OF WAVE BREAKING		BREAKER TYPE	INTENSITY OF ENERGY DISSIPATION
					WAVE GAUGE	WAVE GAUGE LOCATION [m]		
B=1.0m, reef slopes 1:2	h=0.6m	$d_r/h=0.17$	$H_i/h=0.10$	No	-	-	-	-
			$H_i/h=0.17$	No	-	-	-	-
			$H_i/h=0.20$	No	-	-	-	-
			$H_i/h=0.23$	No	-	-	-	-
			$H_i/h=0.27$	Yes	WG17 [v, m]	38.60 LRS	Spilling	Weak
			$H_i/h=0.30$	Yes	WG9 [v]	37.37 BR	Transition	Medium
			$H_i/h=0.33$	Yes	WG8-17 [v]	36.51-38.60 LRS-LRS	Plunging	Weak
			$H_i/h=0.37$	Yes	WG9 v	36.77 LRT	Plunging	Medium
	h=0.7m	$d_r/h=0.43$	$H_i/h=0.31$	No	-	-	-	-
		$d_r/h=0.29$	$H_i/h=0.26$	No	-	-	-	-
			$H_i/h=0.29$	No	-	-	-	-
			$H_i/h=0.31$	No	-	-	-	-

BR – behind reef, LRS – landward reef slope, LRT – landward reef toe, [m] – measurements, [v] – video records
dark grey colour – cases performed in the 2007 experiments

Table C4: Detailed characteristics of solitary wave breaking for trapezoidal reef of width **B=2.0 m** in 2007 and 2008 experiments (H_i=$H_{i,nom}$)

				GENERATION OF WAVE BREAKING	LOCATION OF WAVE BREAKING		BREAKER TYPE	INTENSITY OF ENERGY DISSIPATION
					WAVE GAUGE	WAVE GAUGE LOCATION [m]		
B=2.0m, reef slopes 1:2	h=0.5m	d_r/h=0.4	H_i/h=0.20	No	-	-	-	-
			H_i/h=0.24	No	-	-	-	-
			H_i/h=0.28	No	-	-	-	-
		d_r/h=0.2	H_i/h=0.12	No	-	-	-	-
			H_i/h=0.16	No	-	-	-	-
			H_i/h=0.20	Yes	WG14 ^{v,m}	37.30 LRS	Spilling	Weak
			H_i/h=0.24	Yes	WG8 ^{m}	36.90 LRS	Plunging	Weak
			H_i/h=0.28	Yes	WG21 ^{m}	36.675 RC	Plunging	Medium
	h=0.6m	d_r/h=0.5	H_i/h=0.10	No	-	-	-	-
			H_i/h=0.20	No	-	-	-	-
			H_i/h=0.23	No	-	-	-	-
			H_i/h=0.27	No	-	-	-	-
			H_i/h=0.30	No	-	-	-	-
			H_i/h=0.33	No	-	-	-	-
			H_i/h=0.37	No	-	-	-	-
		d_r/h=0.33	H_i/h=0.10	No	-	-	-	-
			H_i/h=0.17	No	-	-	-	-
			H_i/h=0.20	No	-	-	-	-
			H_i/h=0.23	No	-	-	-	-
			H_i/h=0.27	No	-	-	-	-
			H_i/h=0.30	Yes	WG11 ^{v,m}	38.09 BR	Spilling	Medium
			H_i/h=0.33	Yes	WG16 ^{v,m}	37.75 BR	Spilling	Medium
			H_i/h=0.37	Yes	WG10 ^{v,m}	37.59 LRT	Plunging	Medium
		d_r/h=0.17	H_i/h=0.10	No	-	-	-	-
			H_i/h=0.13	No	-	-	-	-

Table C4: Continuation: Detailed characteristics of solitary wave breaking for trapezoidal reef of width **B=2.0 m** in 2007 and 2008 experiments ($H_i=H_{i,nom}$)

				GENERATION OF WAVE BREAKING	LOCATION OF WAVE BREAKING		BREAKER TYPE	INTENSITY OF ENERGY DISSIPATION
					WAVE GAUGE	WAVE GAUGE LOCATION [m]		
B=2.0m, reef slopes 1:2	h=0.6m	$d_r/h=0.17$	$H_i/h=0.17$	No	-	-	-	-
			$H_i/h=0.20$	Yes	WG10 [v, m]	37.54 LRS	Plunging	Weak
			$H_i/h=0.23$	Yes	WG9-12 [v, m]	37.05-37.175 LRS-LRS	Plunging	Weak
			$H_i/h=0.27$	Yes	WG6-7 [v, m]	36.55-36.80 ARC-LRC	Plunging	Medium
			$H_i/h=0.30$	Yes	WG8 [m]	36.72 LRC	Plunging	Medium
			$H_i/h=0.33$	Yes	WG6-7 [v, m]	36.55-36.80 ARC-LRC	Plunging	Strong
			$H_i/h=0.37$	Yes	WG8 [v, m]	36.72 LRC	Plunging	Strong
	h=0.7m	$d_r/h=0.57$	$H_i/h=0.26$	No	-	-	-	-
			$H_i/h=0.29$	No	-	-	-	-
			$H_i/h=0.31$	No	-	-	-	-
		$d_r/h=0.43$	$H_i/h=0.20$	No	-	-	-	-
			$H_i/h=0.23$	No	-	-	-	-
			$H_i/h=0.26$	No	-	-	-	-
			$H_i/h=0.29$	No	-	-	-	-
			$H_i/h=0.31$	No	-	-	-	-
		$d_r/h=0.29$	$H_i/h=0.14$	No	-	-	-	-
			$H_i/h=0.17$	No	-	-	-	-
			$H_i/h=0.20$	No	-	-	-	-
			$H_i/h=0.23$	No	-	-	-	-
			$H_i/h=0.26$	No	-	-	-	-
			$H_i/h=0.29$	Yes	WG16 [v, m]	38.25 BR	Spilling	Weak
			$H_i/h=0.31$	Yes	WG22 [v,m]	37.95 BR	Spilling	Strong

BR – behind reef, LRS – landward reef slope, LRC – landward reef corner, LRT – landward reef toe, ARC – above reef crown, [m] – measurements, [v] – video records

dark grey colour – cases performed in the 2007 experiments

Appendix D

WAVE FISSION CHARACTERISTICS IN LABORATORY EXPERIMENTS 2007 AND 2008

Table D1: Detailed characteristics of solitary wave fission for trapezoidal reef of width **B=1.0 m** in 2007 and 2008 experiments ($H_i = H_{i,nom}$)

				GENERATION OF WAVE FISSION	BEGINNING OF WAVE FISSION		NUMBER OF SOLITONS	WAVE BREAKING	REMARKS
					WAVE GAUGE	WAVE GAUGE LOCATION [m]			
B=1.0m	$d_r/h=0.5$	With reef slopes 1:2	$H_i/h=0.10$	Yes	WG14-22	40.79-73.79	2	No breaking	Weak distortion
			$H_i/h=0.13$	Yes	WG18-26	38.90-40.79	2	No breaking	-
			$H_i/h=0.17$	Yes	WG15-22	37.90-39.52	2	No breaking	-
			$H_i/h=0.20$	Yes	WG15-22	37.90-39.52	2	No breaking	-
			$H_i/h=0.23$	Yes	WG10-18	37.00-38.90	2	No breaking	-
			$H_i/h=0.27$	Yes	WG9-18	36.40-38.90	2	No breaking	-
			$H_i/h=0.30$	Yes	WG9-16	36.40-38.20	2	No breaking	
			$H_i/h=0.33$	Yes	WG9-16	36.40-38.20	2	No breaking	
			$H_i/h=0.37$	Yes	WG7/8-12	35.77/36.37-38.87	2	No breaking	
	$d_r/h=0.33$	With reef slopes 1:2	$H_i/h=0.10$	Yes	WG12-19	38.37-49.79	3	No breaking	Weak distortion
			$H_i/h=0.13$	Yes	WG12-18	37.20-38.75	3	No breaking	-
			$H_i/h=0.17$	Yes	WG9-16	36.60-38.40	3	No breaking	-
			$H_i/h=0.20$	Yes	WG7/8-11	35.77/36.17-37.77	3	No breaking	-
			$H_i/h=0.23$	Yes	WG9-13	37.20-37.65	3	No breaking	
			$H_i/h=0.27$	Yes	WG9-13	37.20-37.65	3	No breaking	
			$H_i/h=0.30$	Yes	WG9-13	37.20-37.65	3	No breaking	-
			$H_i/h=0.33$	Yes	WG9-13	37.20-37.65	3	No breaking	-
	$d_r/h=0.17$	With reef slopes 1:2	$H_i/h=0.10$	Yes	WG9-14	37.10-37.40	4	No breaking	-
			$H_i/h=0.13$	Yes	WG7-8	36.30-36.80	4	No breaking	-
			$H_i/h=0.17$	Yes	WG6-9	36.80-37.10	4	No breaking	-
			$H_i/h=0.20$	Yes	WG5-6	35.80-36.80	4	No breaking	-
			$H_i/h=0.23$	Yes	WG5-6	35.80-36.80	4	No breaking	-

dark grey colour – results from 2007 experiments repeated in 2008 tests, light grey colour – results from the 2007 experiments

Table D2: Detailed characteristics of solitary wave fission for trapezoidal reef of width **B=2.0 m** in 2007 and 2008 experiments ($H_i=H_{i,nom}$)

				GENERATION OF WAVE FISSION	BEGINNING OF WAVE FISSION		NUMBER OF SOLITONS	WAVE BREAKING	REMARKS
					WAVE GAUGE	WAVE GAUGE LOCATION [m]			
B=2.0m	$d_r/h=0.5$	With reef slopes 1:2	$H_i/h=0.10$	Yes	WG12-20	38.79-53.79	2 (3?)	No breaking	Weak distortion
			$H_i/h=0.13$	Yes	WG13-26	38.30-40.79	2 (3?)	No breaking	-
			$H_i/h=0.17$	Yes	WG12-16	38.00-39.20	3	No breaking	-
			$H_i/h=0.20$	Yes	WG7/8-14	36.09/36.79-39.79	3	No breaking	-
			$H_i/h=0.23$	Yes	WG9-13	37.70-38.30	3	No breaking	-
			$H_i/h=0.27$	Yes	WG8-13	37.40-38.30	3	No breaking	-
			$H_i/h=0.30$	Yes	WG8-13	36.79-39.29	3	No breaking	-
			$H_i/h=0.33$	Yes	WG8-13	37.40-38.30	3	No breaking	-
			$H_i/h=0.37$	Yes	WG8-13	37.40-38.30	3	No breaking	-
	$d_r/h=0.33$	With reef slopes 1:2	$H_i/h=0.10$	Yes	WG12-18	38.20-40.09	3 (4?)	No breaking	Weak distortion
			$H_i/h=0.13$	Yes	WG9-14	37.90-38.80	3 (4?)	No breaking	-
			$H_i/h=0.17$	Yes	WG8-13	37.60-38.50	4	No breaking	-
			$H_i/h=0.20$	Yes	WG9-12	37.19-38.39	4	No breaking	-
			$H_i/h=0.23$	Yes	WG8-12	37.60-38.20	4	No breaking	-
			$H_i/h=0.27$	Yes	WG8-12	37.60-38.20	4	No breaking	-
	$d_r/h=0.17$	With reef slopes 1:2	$H_i/h=0.10$	Yes	WG7-9	37.30-37.80	5	No breaking	-
			$H_i/h=0.13$	Yes	WG6-8	36.80-37.55	5	No breaking	-
			$H_i/h=0.17$	Yes	WG6-8	36.80-37.55	5 (?)	No breaking	-

dark grey colour – results from the 2007 experiments repeated in the 2008 tests, light grey colour – results from the 2007 experiments

Appendix E

COMPARISON OF INCIDENT CONDITIONS OF SOLITARY WAVE AND REEF GEOMETRY USED IN OWN LABORATORY EXPERIMENTS AND NUMERICAL SIMULATIONS

Table E1: Comparison of incident conditions of solitary wave and reef geometry used in own laboratory experiments and numerical simulations (for nominal parameters)

PARAMETER	UNIT	VALUE - EXPERIMENTS	VALUE - SIMULATIONS
h	[m]	0.5, 0.6, 0.7	1.0
d_r/h	[-]	h=0.5: 0.2, 0.4 h=0.6: 0.5, 0.33, 0.17 and 1.0 (no reef) h=0.7: 0.57, 0.43, 0.29	0.3, 0.4, 0.5 and 1.0 (no reef)
h_r/h	[-]	h=0.5: 0.8, 0.6 h=0.6: 0.5, 0.67, 0.83 and 0.0 (no reef) h=0.7: 0.43, 0.57, 0.71	0.7, 0.6, 0.5 and 0.0 (no reef)
B/L_i	[-]	h=0.5: 0.16-0.30 (B=1.0m) 0.33-0.63 (B=2.0m) h=0.6: 0.12-0.24 (B=1.0m) 0.25-0.48 (B=2.0m) h=0.7: 0.13-0.19 (B=1.0m) 0.25-0.38 (B=2.0m)	0.1-1.0 (with increment 0.1) and 0.0 (no reef)
reef slopes	[-]	1:2, no	1:10
H_i/h	[-]	h=0.5: 0.12-0.44 h=0.6: 0.10-0.37 h=0.7: 0.14-0.31	0.1, 0.15, 0.20, 0.25
L_i/h	[-]	h=0.5: 6.4-12.24 h=0.6: 7.0-13.4 h=0.7: 7.6-14.5	100

Appendix F

COMPUTED COEFFICIENTS OF WAVE TRANSMISSION, WAVE REFLECTION AND WAVE ENERGY DISSIPATION

Table F1: Characteristics of processes observed for numerical simulations for H_i/h=0.10, L_i/h=100, h=1.0 m

		WAVE BREAKING		WAVE FISSION		K_t [-]	K_r [-]	K_d [-]	REMARKS
		GENER-ATION	LOCA-TION x/h [-]	GENER-ATION	LOCA-TION x/h [-]				
B/L_i=0	d_r/h=1	No	-	Yes	~95 BBOS (OS)	1.0	0.0	0.0	4 solitons gener. over shelf, 1 soliton fully emerged
B/L_i=0.1	d_r/h=0.5	No	-	Yes	~95 BBOS (OS)	0.979	0.205	0.0	2 solitons gener. in front of reef, many waves gener. behind reef, 1 soliton fully emerged
	d_r/h=0.4	No	-	Yes	~95 BBOS (OS)	0.963	0.270	0.0	2 solitons gener. in front of reef, many waves generated behind reef, 1 soliton fully emerged
	d_r/h=0.3	No	-	Yes	~95 BBOS (OS)	0.938	0.346	0.0	2 solitons gener. in front of reef, many waves gener. behind reef, 1 soliton fully emerged
B/L_i=0.2	d_r/h=0.5	No	-	Yes	~95 BBOS (OS)	0.976	0.220	0.0	2 solitons gener. in front of reef, many waves gener. behind reef, 1 soliton fully emerged
	d_r/h=0.4	No	-	Yes	~95 BBOS (OS)	0.959	0.282	0.0	2 solitons gener. in front of reef, many waves gener. behind reef, 1 soliton fully emerged
	d_r/h=0.3	No	-	Yes	~95 BBOS (OS)	0.931	0.366	0.0	2 solitons gener. in front of reef, many waves gener. behind reef, 1 soliton fully emerged
B/L_i=0.3	d_r/h=0.5	No	-	Yes	~95 BBOS (OS)	0.975	0.220	0.0	2 solitons gener. in front of reef, many waves gener. behind reef, 1 soliton fully emerged
	d_r/h=0.4	No	-	Yes	~95 BBOS (OS)	0.958	0.285	0.0	2 solitons gener. in front of reef, many waves gener. behind reef, almost 1 soliton fully emerged
	d_r/h=0.3	No	-	Yes	~95 BBOS (OS)	0.921	0.390	0.0	2 solitons gener. in front of reef, many waves gener. behind reef, almost 1 soliton fully emerged
B/L_i=0.4	d_r/h=0.5	No	-	Yes	~95 BBOS (OS)	0.976	0.219	0.0	2 solitons gener. in front of reef, many waves gener. behind reef, almost 1 soliton fully emerged
	d_r/h=0.4	No	-	Yes	~95 BBOS (OS)	0.957	0.290	0.0	2 solitons gener. in front of reef, many waves gener. behind reef, almost 1 soliton fully emerged
	d_r/h=0.3	No	-	Yes	~95 BBOS (OS)	0.915	0.402	0.0	2 solitons gener. in front of reef, many waves gener.behind reef, almost 1 soliton fully emerged

Table F1: Continuation: Characteristics of processes observed for numerical simulations for H_i/h=0.10, L_i/h=100, h=1.0 m

		WAVE BREAKING		WAVE FISSION		K_t [-]	K_r [-]	K_d [-]	REMARKS
		GENER-ATION	LOCA-TION x/h [-]	GENER-ATION	LOCA-TION x/h [-]				
B/L_i=0.5	d_r/h=0.5	No	-	Yes	~95 BBOS (OS)	0.976	0.219	0.0	2 solitons gener. in front of reef, many waves gener. behind reef, almost 1 soliton fully emerged
	d_r/h=0.4	No	-	Yes	~95 BBOS (OS)	0.957	0.291	0.0	2 solitons gener. in front of reef, many waves gener. behind reef, almost 1 soliton fully emerged
	d_r/h=0.3	No	-	Yes	~95 BBOS (OS)	0.913	0.407	0.0	2 solitons gener. in front of reef, many waves gener. behind reef, almost 1 soliton fully emerged
B/L_i=0.6	d_r/h=0.5	No	-	Yes	~95 BBOS (OS)	0.976	0.219	0.0	2 solitons gener. in front of reef, many waves gener. behind reef, almost 1 soliton fully emerged
	d_r/h=0.4	No	-	Yes	~95 BBOS (OS)	0.956	0.293	0.0	2 solitons gener. in front of reef, many waves gener. behind reef, almost 1 soliton fully emerged
	d_r/h=0.3	No	-	Yes	~95 BBOS (OS)	0.911	0.412	0.0	2 solitons gener. in front of reef, many waves gener. behind reef, no soliton fully emerged
B/L_i=0.7	d_r/h=0.5	No	-	Yes	~95 BBOS (OS)	0.975	0.220	0.0	2 solitons gener. in front of reef, many waves gener. behind reef, almost 1 soliton fully emerged
	d_r/h=0.4	No	-	Yes	~95 BBOS (OS)	0.956	0.294	0.0	2 solitons gener. in front of reef, many waves gener. behind reef, almost 1 soliton fully emerged
	d_r/h=0.3	No	-	Yes	~95 BBOS (OS)	0.911	0.413	0.0	2 solitons gener. in front of reef, many waves gener. behind reef, no soliton fully emerged
B/L_i=0.8	d_r/h=0.5	No	-	Yes	~95 BBOS (OS)	0.975	0.221	0.0	2 solitons gener. in front of reef, many waves gener. behind reef, almost 1 soliton fully emerged
	d_r/h=0.4	No	-	Yes	~95 BBOS (OS)	0.955	0.295	0.0	2 solitons gener. in front of reef, many waves gener. behind reef, almost 1 soliton fully emerged
	d_r/h=0.3	No	-	Yes	~95 BBOS (OS)	0.912	0.409	0.0	2 solitons gener. in front of reef, many waves gener. behind reef, no soliton fully emerged

Table F1: Continuation: Characteristics of processes observed for numerical simulations for $H_i/h=0.10$, $L_i/h=100$, $h=1.0$ m

		WAVE BREAKING		WAVE FISSION		K_t [-]	K_r [-]	K_d [-]	REMARKS
		GENER-ATION	LOCA-TION x/h [-]	GENER-ATION	LOCA-TION x/h [-]				
$B/L_i=0.9$	$d_r/h=0.5$	No	-	Yes	~95 BBOS (OS)	0.975	0.221	0.0	2 solitons gener. in front of reef, many waves gener. behind reef, no soliton fully emerged
	$d_r/h=0.4$	No	-	Yes	~95 BBOS (OS)	0.956	0.295	0.0	2 solitons gener. in front of reef, many waves gener. behind reef, no soliton fully emerged
	$d_r/h=0.3$	No	-	Yes	~95 BBOS (OS)	0.914	0.405	0.0	2 solitons gener. in front of reef, many waves gener. behind reef, no soliton fully emerged
$B/L_i=1.0$	$d_r/h=0.5$	No	-	Yes	~95 BBOS (OS)	0.975	0.221	0.0	2 solitons gener. in front of reef, many waves gener. behind reef, no soliton fully emerged
	$d_r/h=0.4$	No	-	Yes	~95 BBOS (OS)	0.956	0.294	0.0	2 solitons gener. in front of reef, many waves gener. behind reef, no soliton fully emerged
	$d_r/h=0.3$	No	-	Yes	~95 BBOS (OS)	0.915	0.402	0.0	2 solitons gener. in front of reef, many waves gener. behind reef, no soliton fully emerged

BBOS – behind beginning of shelf, OS – over shelfbeginning of horizontal part of shelf, BSRT – behind seaward reef toe, ORC – over reef crest, OS – over shelf, gener. – generated

Table F2: Characteristics of processes observed for numerical simulations for $H_i/h=0.15$, $L_i/h=100$, $h=1.0$ m

		WAVE BREAKING		WAVE FISSION					
		GENER-ATION	LOCA-TION x/h [-]	GENER-ATION	LOCA-TION x/h [-]	K_t [-]	K_r [-]	K_d [-]	REMARKS
$B/L_i=0$	$d_r/h=1$	No	-	Yes	~57 BBOS (OS)	1.0	0.0	0.0	5 solitons gener. over shelf, 2 solitons fully emerged
$B/L_i=0.1$	$d_r/h=0.5$	No	-	Yes	~57 BBOS (OS)	0.981	0.196	0.0	2 solitons gener. in front of reef, many waves gener. behind reef, 2 solitons fully emerged
	$d_r/h=0.4$	No	-	Yes	~57 BBOS (OS)	0.965	0.263	0.0	2 solitons gener. in front of reef, many waves generated behind reef, 2 solitons fully emerged
	$d_r/h=0.3$	No	-	Yes	~57 BBOS (OS)	0.935	0.355	0.0	2 solitons gener. in front of reef, many waves gener. behind reef, 2 solitons fully emerged
$B/L_i=0.2$	$d_r/h=0.5$	No	-	Yes	~57 BBOS (OS)	0.976	0.216	0.0	2 solitons gener. in front of reef, many waves gener. behind reef, 2 solitons fully emerged
	$d_r/h=0.4$	No	-	Yes	~57 BBOS (OS)	0.958	0.288	0.0	2 solitons gener. in front of reef, many waves gener. behind reef, 2 solitons fully emerged
	$d_r/h=0.3$	Yes	~21 BSRT (ORC)	Yes	~57 BBOS (OS)	0.890	0.359	0.282	2 solitons gener. in front of reef, many waves gener. behind reef, 2 solitons fully emerged
$B/L_i=0.3$	$d_r/h=0.5$	No	-	Yes	~57 BBOS (OS)	0.976	0.217	0.0	2 solitons gener. in front of reef, many waves gener. behind reef, 2 solitons fully emerged
	$d_r/h=0.4$	No	-	Yes	~57 BBOS (OS)	0.957	0.290	0.0	2 solitons gener. in front of reef, many waves gener. behind reef, 1 soliton fully emerged
	$d_r/h=0.3$	Yes	~21 BSRT (ORC)	Yes	~57 BBOS (OS)	0.869	0.357	0.343	2 solitons gener. in front of reef, many waves gener. behind reef, 1 soliton fully emerged
$B/L_i=0.4$	$d_r/h=0.5$	No	-	Yes	~57 BBOS (OS)	0.976	0.216	0.0	2 solitons gener. in front of reef, many waves gener. behind reef, 1 soliton fully emerged
	$d_r/h=0.4$	No	-	Yes	~57 BBOS (OS)	0.957	0.291	0.0	2 solitons gener. in front of reef, many waves gener. behind reef, 1 soliton fully emerged
	$d_r/h=0.3$	Yes	~21 BSRT (ORC)	Yes	~57 BBOS (OS)	0.853	0.354	0.382	2 solitons gener. in front of reef, many waves gener.behind reef, 1 soliton fully emerged

Table F2: Continuation: Characteristics of processes observed for numerical simulations for $H_i/h=0.15$, $L_i/h=100$, $h=1.0$ m

		WAVE BREAKING		WAVE FISSION					
		GENER-ATION	LOCA-TION x/h [-]	GENER-ATION	LOCA-TION x/h [-]	K_t [-]	K_r [-]	K_d [-]	REMARKS
$B/L_i=0.5$	$d_r/h=0.5$	No	-	Yes	~57 BBOS (OS)	0.976	0.216	0.0	2 solitons gener. in front of reef, many waves gener. behind reef, 1 soliton fully emerged
	$d_r/h=0.4$	No	-	Yes	~57 BBOS (OS)	0.956	0.294	0.0	2 solitons gener. in front of reef, many waves gener. behind reef, 1 soliton fully emerged
	$d_r/h=0.3$	Yes	~21 BSRT (ORC)	Yes	~57 BBOS (OS)	0.837	0.354	0.418	2 solitons gener. in front of reef, many waves gener. behind reef, 1 soliton fully emerged
$B/L_i=0.6$	$d_r/h=0.5$	No	-	Yes	~57 BBOS (OS)	0.976	0.218	0.0	2 solitons gener. in front of reef, many waves gener. behind reef, no soliton fully emerged
	$d_r/h=0.4$	No	-	Yes	~57 BBOS (OS)	0.954	0.301	0.0	2 solitons gener. in front of reef, many waves gener. behind reef, no soliton fully emerged
	$d_r/h=0.3$	Yes	~21 BSRT (ORC)	Yes	~57 BBOS (OS)	0.822	0.351	0.448	2 solitons gener. in front of reef, many waves gener. behind reef, 1 soliton fully emerged
$B/L_i=0.7$	$d_r/h=0.5$	No	-	Yes	~57 BBOS (OS)	0.976	0.219	0.0	2 solitons gener. in front of reef, many waves gener. behind reef, no soliton fully emerged
	$d_r/h=0.4$	No	-	Yes	~57 BBOS (OS)	0.953	0.302	0.0	2 solitons gener. in front of reef, many waves gener. behind reef, no soliton fully emerged
	$d_r/h=0.3$	Yes	~21 BSRT (ORC)	Yes	~57 BBOS (OS)	0.809	0.348	0.473	2 solitons gener. in front of reef, many waves gener. behind reef, 1 soliton fully emerged
$B/L_i=0.8$	$d_r/h=0.5$	No	-	Yes	~57 BBOS (OS)	0.976	0.220	0.0	2 solitons gener. in front of reef, many waves gener. behind reef, 1 soliton fully emerged
	$d_r/h=0.4$	No	-	Yes	~57 BBOS (OS)	0.953	0.304	0.0	2 solitons gener. in front of reef, many waves gener. behind reef, no soliton fully emerged
	$d_r/h=0.3$	Yes	~21 BSRT (ORC)	Yes	~57 BBOS (OS)	0.795	0.346	0.498	2 solitons gener. in front of reef, many waves gener. behind reef, 1 soliton fully emerged

Table F2: Continuation: Characteristics of processes observed for numerical simulations for $H_i/h=0.15$, $L_i/h=100$, $h=1.0$ m

		WAVE BREAKING		WAVE FISSION		K_t [-]	K_r [-]	K_d [-]	REMARKS
		GENER-ATION	LOCA-TION x/h [-]	GENER-ATION	LOCA-TION x/h [-]				
$B/L_i=0.9$	**$d_r/h=0.5$**	No	-	Yes	~57 BBOS (OS)	0.975	0.220	0.0	2 solitons gener. in front of reef, many waves gener. behind reef, 1 soliton fully emerged
	$d_r/h=0.4$	No	-	Yes	~57 BBOS (OS)	0.954	0.299	0.0	2 solitons gener. in front of reef, many waves gener. behind reef, no soliton fully emerged
	$d_r/h=0.3$	Yes	~21 BSRT (ORC)	Yes	~57 BBOS (OS)	0.781	0.344	0.522	2 solitons gener. in front of reef, many waves gener. behind reef, almost 1 soliton fully emerged
$B/L_i=1.0$	**$d_r/h=0.5$**	No	-	Yes	~57 BBOS (OS)	0.976	0.220	0.0	2 solitons gener. in front of reef, many waves gener. behind reef, almost 1 soliton fully emerged
	$d_r/h=0.4$	No	-	Yes	~57 BBOS (OS)	0.953	0.303	0.0	2 solitons gener. in front of reef, many waves gener. behind reef, almost 1 soliton fully emerged
	$d_r/h=0.3$	Yes	~21 BSRT (ORC)	Yes	~57 BBOS (OS)	0.769	0.344	0.539	2 solitons gener. in front of reef, many waves gener. behind reef, almost 1 soliton fully emerged

BBOS – behind beginning of shelf, BSRT – behind seaward reef toe, ORC – over reef crest, OS – over shelfbeginning S – behind beginning of shelf, BSRT – behind seaward reef toe, ORC – over reef crest, OS – over shelf, gener. – generated

Table F3: Characteristics of processes observed for numerical simulations for $H_i/h=0.20$, $L_i/h=100$, $h=1.0$ m

		WAVE BREAKING		WAVE FISSION		K_t [-]	K_r [-]	K_d [-]	REMARKS
		GENER-ATION	LOCA-TION x/h [-]	GENER-ATION	LOCA-TION x/h [-]				
$B/L_i=0$	$d_r/h=1$	No	-	Yes	~35 BBOS (OS)	1.0	0.0	0.0	6 solitons gener. over shelf, 3 solitons fully emerged
$B/L_i=0.1$	$d_r/h=0.5$	No	-	Yes	~35 BBOS (OS)	0.982	0.190	0.0	3 solitons gener. in front of reef, many waves gener. behind reef, 3 solitons fully emerged
	$d_r/h=0.4$	No	-	Yes	~35 BBOS (OS)	0.965	0.261	0.0	3 solitons gener. in front of reef, many waves generated behind reef, 3 solitons fully emerged
	$d_r/h=0.3$	Yes	~13 BSRT (ORC)	Yes	~35 BBOS (OS)	0.901	0.331	0.280	3 solitons gener. in front of reef, many waves gener. behind reef, 3 solitons fully emerged
$B/L_i=0.2$	$d_r/h=0.5$	No	-	Yes	~35 BBOS (OS)	0.978	0.209	0.0	3 solitons gener. in front of reef, many waves gener. behind reef, 3 solitons fully emerged
	$d_r/h=0.4$	Yes	~20 BSRT (ORC)	Yes	~35 BBOS (OS)	0.917	0.275	0.288	3 solitons gener. in front of reef, many waves gener. behind reef, 3 solitons fully emerged
	$d_r/h=0.3$	Yes	~13 BSRT (ORC)	Yes	~35 BBOS (OS)	0.871	0.336	0.358	3 solitons gener. in front of reef, many waves gener. behind reef, 3 solitons fully emerged
$B/L_i=0.3$	$d_r/h=0.5$	No	-	Yes	~35 BBOS (OS)	0.978	0.209	0.0	3 solitons gener. in front of reef, many waves gener. behind reef, 3 solitons fully emerged
	$d_r/h=0.4$	Yes	~20 BSRT (ORC)	Yes	~35 BBOS (OS)	0.894	0.277	0.353	3 solitons gener. in front of reef, many waves gener. behind reef, 3 solitons fully emerged
	$d_r/h=0.3$	Yes	~13 BSRT (ORC)	Yes	~35 BBOS (OS)	0.852	0.336	0.402	3 solitons gener. in front of reef, many waves gener. behind reef, 3 solitons fully emerged
$B/L_i=0.4$	$d_r/h=0.5$	No	-	Yes	~35 BBOS (OS)	0.977	0.214	0.0	3 solitons gener. in front of reef, many waves gener. behind reef, 3 solitons fully emerged
	$d_r/h=0.4$	Yes	~20 BSRT (ORC)	Yes	~35 BBOS (OS)	0.880	0.272	0.390	3 solitons gener. in front of reef, many waves gener. behind reef, 2 solitons fully emerged
	$d_r/h=0.3$	Yes	~13 BSRT (ORC)	Yes	~35 BBOS (OS)	0.837	0.334	0.434	3 solitons gener. in front of reef, many waves gener.behind reef, 2 solitons fully emerged

Table F3: Continuation: Characteristics of processes observed for numerical simulations for H_i/h=0.20, L_i/h=100, h=1.0 m

		WAVE BREAKING		WAVE FISSION		K_t [-]	K_r [-]	K_d [-]	REMARKS
		GENER-ATION	LOCA-TION x/h [-]	GENER-ATION	LOCA-TION x/h [-]				
B/L_i=0.5	**d_r/h=0.5**	No	-	Yes	~35 BBOS (OS)	0.977	0.215	0.0	3 solitons gener. in front of reef, many waves gener. behind reef, 3 solitons fully emerged
	d_r/h=0.4	Yes	~20 BSRT (ORC)	Yes	~35 BBOS (OS)	0.869	0.272	0.413	3 solitons gener. in front of reef, many waves gener. behind reef, 2 solitons fully emerged
	d_r/h=0.3	Yes	~13 BSRT (ORC)	Yes	~35 BBOS (OS)	0.823	0.330	0.461	3 solitons gener. in front of reef, many waves gener. behind reef, 2 solitons fully emerged
B/L_i=0.6	**d_r/h=0.5**	No	-	Yes	~35 BBOS (OS)	0.975	0.224	0.0	3 solitons gener. in front of reef, many waves gener. behind reef, 2 solitons fully emerged
	d_r/h=0.4	Yes	~20 BSRT (ORC)	Yes	~35 BBOS (OS)	0.861	0.271	0.431	3 solitons gener. in front of reef, many waves gener. behind reef, 2 solitons fully emerged
	d_r/h=0.3	Yes	~13 BSRT (ORC)	Yes	~35 BBOS (OS)	0.807	0.329	0.491	3 solitons gener. in front of reef, many waves gener. behind reef, 1 soliton fully emerged
B/L_i=0.7	**d_r/h=0.5**	No	-	Yes	~35 BBOS (OS)	0.976	0.216	0.0	3 solitons gener. in front of reef, many waves gener. behind reef, 2 solitons fully emerged
	d_r/h=0.4	Yes	~20 BSRT (ORC)	Yes	~35 BBOS (OS)	0.853	0.268	0.447	3 solitons gener. in front of reef, many waves gener. behind reef, almost 2 solitons fully emerged
	d_r/h=0.3	Yes	~13 BSRT (ORC)	Yes	~35 BBOS (OS)	0.788	0.328	0.522	3 solitons gener. in front of reef, many waves gener. behind reef, 1 soliton fully emerged
B/L_i=0.8	**d_r/h=0.5**	No	-	Yes	~35 BBOS (OS)	0.977	0.215	0.0	3 solitons gener. in front of reef, many waves gener. behind reef, 2 solitons fully emerged
	d_r/h=0.4	Yes	~20 BSRT (ORC)	Yes	~35 BBOS (OS)	0.842	0.265	0.470	3 solitons gener. in front of reef, many waves gener. behind reef, almost 2 solitons fully emerged
	d_r/h=0.3	Yes	~13 BSRT (ORC)	Yes	~35 BBOS (OS)	0.773	0.325	0.544	3 solitons gener. in front of reef, many waves gener. behind reef, 1 soliton fully emerged

Table F3: Continuation: Characteristics of processes observed for numerical simulations for H_i/h=0.20, L_i/h=100, h=1.0 m

		WAVE BREAKING		WAVE FISSION		K_t [-]	K_r [-]	K_d [-]	REMARKS
		GENER-ATION	LOCA-TION x/h [-]	GENER-ATION	LOCA-TION x/h [-]				
B/L_i=0.9	d_r/h=0.5	No	-	Yes	~35 BBOS (OS)	0.976	0.217	0.0	3 solitons gener. in front of reef, many waves gener. behind reef, almost 2 solitons fully emerged
	d_r/h=0.4	Yes	~20 BSRT (ORC)	Yes	~35 BBOS (OS)	0.829	0.263	0.494	3 solitons gener. in front of reef, many waves gener. behind reef, almost 2 solitons fully emerged
	d_r/h=0.3	Yes	~13 BSRT (ORC)	Yes	~35 BBOS (OS)	0.763	0.325	0.559	3 solitons gener. in front of reef, many waves gener. behind reef, 1 soliton fully emerged
B/L_i=1.0	d_r/h=0.5	No	-	Yes	~35 BBOS (OS)	0.976	0.218	0.0	3 solitons gener. in front of reef, many waves gener. behind reef, 1 soliton fully emerged
	d_r/h=0.4	Yes	~20 BSRT (ORC)	Yes	~35 BBOS (OS)	0.816	0.262	0.514	3 solitons gener. in front of reef, many waves gener. behind reef, 1 soliton fully emerged
	d_r/h=0.3	Yes	~13 BSRT (ORC)	Yes	~35 BBOS (OS)	0.753	0.325	0.572	3 solitons gener. in front of reef, many waves gener. behind reef, 1 soliton fully emerged

BBOS – behind beginning of shelf, BSRT – behind seaward reef toe, ORC – over reef crest, OS – over shelfbeginning BOS – behind beginning of shelf, BSRT – behind seaward reef toe, ORC – over reef crest, OS – over shelf, gener. – generated

Table F4: Characteristics of processes observed for numerical simulations for $H_i/h=0.25$, $L_i/h=100$, $h=1.0$ m

		WAVE BREAKING		WAVE FISSION					
		GENER-ATION	LOCA-TION x/h [-]	GENER-ATION	LOCA-TION x/h [-]	K_t [-]	K_r [-]	K_d [-]	REMARKS
$B/L_i=0$	$d_r/h=1$	No	-	Yes	~26 BBOS (OS)	1.0	0.0	0.0	7 solitons gener. over shelf, 4 solitons fully emerged
$B/L_i=0.1$	$d_r/h=0.5$	No	-	Yes	~26 BBOS (OS)	0.983	0.183	0.0	3 solitons gener. in front of reef, many waves gener. behind reef, 4 solitons fully emerged
	$d_r/h=0.4$	Yes	~12 BSRT (ORC)	Yes	~26 BBOS (OS)	0.929	0.251	0.274	3 solitons gener. in front of reef, many waves gener. behind reef, 4 solitons fully emerged
	$d_r/h=0.3$	Yes	~10 BSRT (ORC)	Yes	~26 BBOS (OS)	0.892	0.316	0.324	3 solitons gener. in front of reef, many waves gener. behind reef, 4 solitons fully emerged
$B/L_i=0.2$	$d_r/h=0.5$	Yes	~23 BSRT (ORC)	Yes	~26 BBOS (OS)	0.955	0.211	0.209	3 solitons gener. in front of reef, many waves gener. behind reef, 4 solitons fully emerged
	$d_r/h=0.4$	Yes	~12 BSRT (ORC)	Yes	~26 BBOS (OS)	0.898	0.264	0.352	3 solitons gener. in front of reef, many waves gener. behind reef, 4 solitons fully emerged
	$d_r/h=0.3$	Yes	~10 BSRT (ORC)	Yes	~26 BBOS (OS)	0.864	0.324	0.386	3 solitons gener. in front of reef, many waves gener. behind reef, 4 solitons fully emerged
$B/L_i=0.3$	$d_r/h=0.5$	Yes	~23 BSRT (ORC)	Yes	~26 BBOS (OS)	0.920	0.214	0.328	3 solitons gener. in front of reef, many waves gener. behind reef, 4 solitons fully emerged
	$d_r/h=0.4$	Yes	~12 BSRT (ORC)	Yes	~26 BBOS (OS)	0.880	0.263	0.395	3 solitons gener. in front of reef, many waves gener. behind reef, 4 solitons fully emerged
	$d_r/h=0.3$	Yes	~10 BSRT (ORC)	Yes	~26 BBOS (OS)	0.847	0.324	0.422	3 solitons gener. in front of reef, many waves gener. behind reef, 4 solitons fully emerged
$B/L_i=0.4$	$d_r/h=0.5$	Yes	~23 BSRT (ORC)	Yes	~26 BBOS (OS)	0.904	0.210	0.373	3 solitons gener. in front of reef, many waves gener. behind reef, 4 solitons fully emerged
	$d_r/h=0.4$	Yes	~12 BSRT (ORC)	Yes	~26 BBOS (OS)	0.868	0.261	0.422	3 solitons gener. in front of reef, many waves gener. behind reef, 4 solitons fully emerged
	$d_r/h=0.3$	Yes	~10 BSRT (ORC)	Yes	~26 BBOS (OS)	0.835	0.319	0.449	3 solitons gener. in front of reef, many waves gener.behind reef, 3 solitons fully emerged

Table F4: Continuation: Characteristics of processes observed for numerical simulations for $H_i/h=0.25$, $L_i/h=100$, $h=1.0$ m

		WAVE BREAKING		WAVE FISSION					
		GENER-ATION	LOCA-TION x/h [-]	GENER-ATION	LOCA-TION x/h [-]	K_t [-]	K_r [-]	K_d [-]	REMARKS
$B/L_i=0.5$	**$d_r/h=0.5$**	Yes	~23 BSRT (ORC)	Yes	~26 BBOS (OS)	0.891	0.208	0.403	3 solitons gener. in front of reef, many waves gener. behind reef, 2 solitons fully emerged
	$d_r/h=0.4$	Yes	~12 BSRT (ORC)	Yes	~26 BBOS (OS)	0.858	0.262	0.441	3 solitons gener. in front of reef, many waves gener. behind reef, 2 solitons fully emerged
	$d_r/h=0.3$	Yes	~10 BSRT (ORC)	Yes	~26 BBOS (OS)	0.818	0.317	0.480	3 solitons gener. in front of reef, many waves gener. behind reef, 3 solitons fully emerged
$B/L_i=0.6$	**$d_r/h=0.5$**	Yes	~23 BSRT (ORC)	Yes	~26 BBOS (OS)	0.882	0.205	0.424	3 solitons gener. in front of reef, many waves gener. behind reef, 2 solitons fully emerged
	$d_r/h=0.4$	Yes	~12 BSRT (ORC)	Yes	~26 BBOS (OS)	0.850	0.258	0.458	3 solitons gener. in front of reef, many waves gener. behind reef, almost 3 solitons fully emerged
	$d_r/h=0.3$	Yes	~10 BSRT (ORC)	Yes	~26 BBOS (OS)	0.795	0.316	0.517	3 solitons gener. in front of reef, many waves gener. behind reef, almost 3 solitons fully emerged
$B/L_i=0.7$	**$d_r/h=0.5$**	Yes	~23 BSRT (ORC)	Yes	~26 BBOS (OS)	0.875	0.202	0.440	3 solitons gener. in front of reef, many waves gener. behind reef, 2 solitons fully emerged
	$d_r/h=0.4$	Yes	~12 BSRT (ORC)	Yes	~26 BBOS (OS)	0.835	0.257	0.486	3 solitons gener. in front of reef, many waves gener. behind reef, almost 3 solitons fully emerged
	$d_r/h=0.3$	Yes	~10 BSRT (ORC)	Yes	~26 BBOS (OS)	0.779	0.314	0.542	3 solitons gener. in front of reef, many waves gener. behind reef, almost 2 solitons fully emerged
$B/L_i=0.8$	**$d_r/h=0.5$**	Yes	~23 BSRT (ORC)	Yes	~26 BBOS (OS)	0.870	0.205	0.449	3 solitons gener. in front of reef, many waves gener. behind reef, almost 3 solitons fully emerged
	$d_r/h=0.4$	Yes	~12 BSRT (ORC)	Yes	~26 BBOS (OS)	0.819	0.256	0.514	3 solitons gener. in front of reef, many waves gener. behind reef, 1 soliton fully emerged
	$d_r/h=0.3$	Yes	~10 BSRT (ORC)	Yes	~26 BBOS (OS)	0.768	0.314	0.557	3 solitons gener. in front of reef, many waves gener. behind reef, 4 solitons fully emerged

BBOS – behind beginning of shelf, BSRT – behind seaward reef toe, ORC – over reef crest, OS – over shelf, gener. - generated

Table F4: Continuation: Characteristics of processes observed for numerical simulations for $H_i/h=0.25$, $L_i/h=100$, $h=1.0$ m

		WAVE BREAKING		WAVE FISSION					
		GENER-ATION	LOCA-TION x/h [-]	GENER-ATION	LOCA-TION x/h [-]	K_t [-]	K_r [-]	K_d [-]	REMARKS
$B/L_i=0.9$	$d_r/h=0.5$	Yes	~23 BSRT (ORC)	Yes	~26 BBOS (OS)	0.865	0.203	0.459	3 solitons gener. in front of reef, many waves gener. behind reef, almost 3 solitons fully emerged
	$d_r/h=0.4$	Yes	~12 BSRT (ORC)	Yes	~26 BBOS (OS)	0.806	0.255	0.534	3 solitons gener. in front of reef, many waves gener. behind reef, 1 soliton fully emerged
	$d_r/h=0.3$	Yes	~10 BSRT (ORC)	Yes	~26 BBOS (OS)	0.760	0.313	0.570	3 solitons gener. in front of reef, many waves gener. behind reef, 1 soliton fully emerged
$B/L_i=1.0$	$d_r/h=0.5$	Yes	~23 BSRT (ORC)	Yes	~26 BBOS (OS)	0.858	0.201	0.473	3 solitons gener. in front of reef, many waves gener. behind reef, 2 solitons fully emerged
	$d_r/h=0.4$	Yes	~12 BSRT (ORC)	Yes	~26 BBOS (OS)	0.796	0.252	0.550	3 solitons gener. in front of reef, many waves gener. behind reef, 1 soliton fully emerged
	$d_r/h=0.3$	Yes	~10 BSRT (ORC)	Yes	~26 BBOS (OS)	0.747	0.312	0.586	3 solitons gener. in front of reef, many waves gener. behind reef, 1 soliton fully emerged

BBOS – behind beginning of shelf, BSRT – behind seaward reef toe, ORC – over reef crest, OS – over shelfbeginning beginning of shelf, BSRT – behind seaward reef toe, ORC – over reef crest, OS – over shelf, gener. - generated

Appendix G

RESULTS OF NUMERICAL STUDY OF LIN (2004) ON FISSION OF NONBREAKING AND BREAKING SOLITARY WAVES OVER SUBMERGED STRUCTURE OF FINITE WIDTH

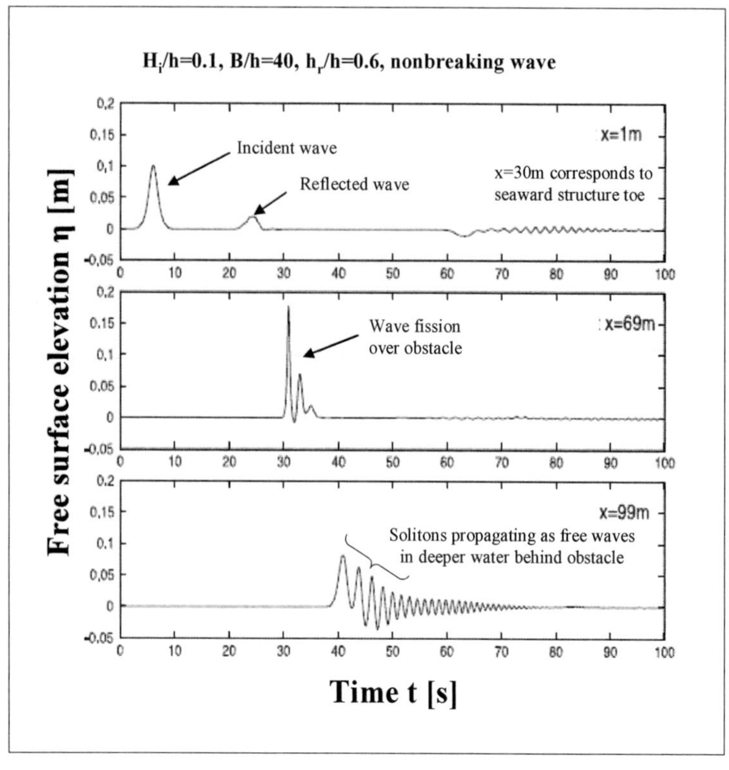

Figure G1: Generation of fission for nonbreaking solitary wave in investigation of Lin (2004)

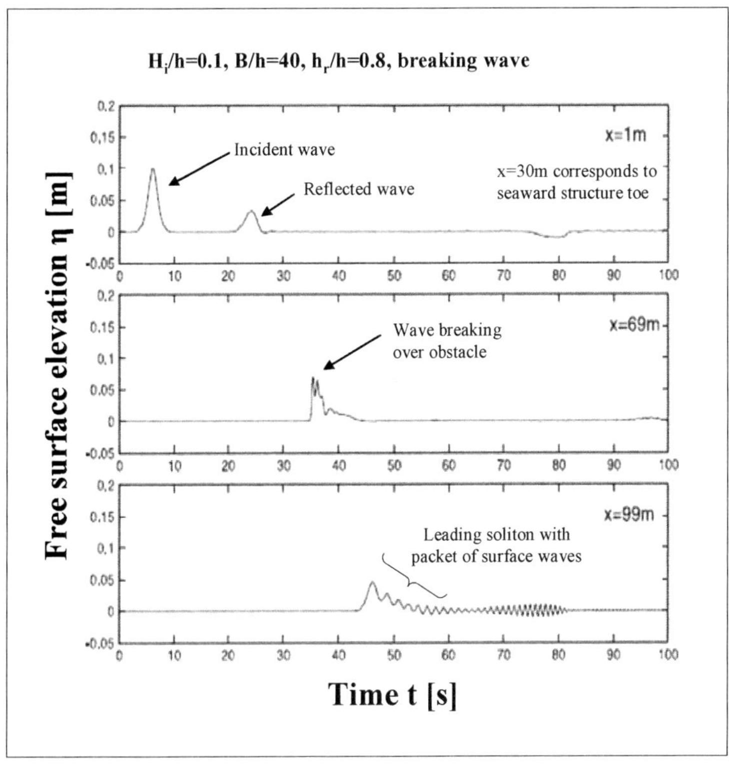

Figure G2: Generation of fission for breaking solitary wave in investigation of Lin (2004)

ibidem-Verlag
Melchiorstr. 15
D-70439 Stuttgart
info@ibidem-verlag.de

www.ibidem-verlag.de
www.ibidem.eu
www.edition-noema.de
www.autorenbetreuung.de